e-Patients Live Longer
Managing Healthcare Using Technology

by Nancy B. Finn M. Ed

Foreword by Daniel Z Sands, MD, MPH Harvard Medical

DORRANCE
PUBLISHING CO
EST. 1920
PITTSBURGH, PENNSYLVANIA 15238

Dorrance Publishing Co
585 Alpha Drive
Suite 103
Pittsburgh, PA 15238
Visit our website at *www.dorrancebookstore.com*

Library of Congress Control Number: 2011910363

ISBN: 979-8-88683-354-6
EISBN: 979-8-88683-332-4

e-Patients Live Longer

Managing Healthcare Using Technology

DEDICATION

This book is about empowering every patient through technology and health applications. It is dedicated to Danny Sands MD, MPH, and to the Society for Participatory Medicine an organization that has been working for several years to move the needle in healthcare to a place where digital technology that empowers e-Patients is fully integrated into the healthcare landscape by both patients and providers working as a team to achieve coordinated, participatory care for all patients.

Daniel Z. Sands, MD, MPH, is an Assistant Clinical Professor of Medicine, at Harvard Medical School, and a staff physician at Beth Israel Deaconess Medical Center, Co-Founder and Chair Emeritus, of the Society for Participatory Medicine. Danny was among the first physicians in medicine to realize and promote the value of digital health records and has been responsible for raising the awareness of all healthcare consumers regarding how important it is that they be a participatory team member in managing and monitoring their own health.

The Society for Participatory Medicine (SPM) was created in 2009 by a passionate group of healthcare professionals, patients, and caregivers with a vision to start a global movement to transform the culture of healthcare. SPM is a 501(c)(3) not-for-profit membership organization. SPM's mission is for all patients, caregivers, and healthcare professionals to fully participate in a collaborative journey to better physical, emotional, and social health.

FOREWORD

"…communication extends all the way from the intimate interaction between a clinician and a patient, to the most public dissemination of information."

Harvey Fienberg, MD, PhD
President, Gordon and Betty Moore Foundation
Former President, Institute of Medicine

Communication has been and always will be at the heart of healthcare. For most of the profession's history, physicians could do very little useful for patients but listen to them and offer them advice. Yet physicians who communicated effectively and compassionately well were still generally revered.

These same communication skills are arguably more important today, despite (or perhaps because of) the proliferation of technologies physicians can effectively apply to alleviate pain and suffering and prolong life. And yet, physicians often neglect these skills, due to the pressures of seeing ever more patients in decreasing amounts of time. Because of these pressures, physicians have increasingly focused on regurgitating facts and ordering tests and consultations, rather than spending time talking and listening to their patients.

But communication with patients is critical to patient engagement. Patient engagement in the care process results in greater satisfaction, better adherence to care plans, improved outcomes, and consequently reduced healthcare costs. Although these costs savings do not accrue to the physicians in a fee-for-service world, they could ultimately result in improved reimbursements (especially true as we move to value-based payment models). So ironically, the same financial pressures that are preventing physicians from taking the time to listen to their patients could be alleviated through better communication with their patients. In other words, investing time in connecting with patients pays dividends.

But the quality of communication is just one aspect. The tools we use to communicate are also important. Historically, all communication between patients and physicians took place in person. But with the advent of the telephone in the early twentieth century, physicians and patients learned how to best apply this technology for the betterment of patient care. Today there are many other ways we can connect with patients, including electronic messaging of various types, video conferencing/telemedicine, and even social networking. But physicians have been reluctant to adopt these technologies, despite their apparent benefits, for a variety of reasons (some legitimate and many not). Even relatively low-tech tools like email, which burst on the popular scene with the advent of the World Wide Web in the mid-1990s, is still shunned by the majority of physicians, despite guidelines that were published over two decades ago and regulations encouraging its use.

The COVID-19 pandemic highlighted the importance of these alternative care channels. Many medical offices were shut down and patients would not, or could not, leave their homes safely, and yet patients still required care. Payment for non-visit-based care was instituted and practices had to learn how to take care of patients who were not in the office, using telephone calls, electronic messaging, and synchronous video interactions. Both physicians and their patients needed to learn how these technologies could be used to best care for patients, including how to use data generated by patients as a data source. Although there were patients—such as those lacking access to reliable technology, who were unable to find private areas in their homes, or lacked the literacy or language skills to stay connected to care—many, including those with limited mobility or inability to take time off from work, benefited greatly from this new type of care delivery. These visits became virtual house calls, and often permitted a greater frequency of touchpoints and greater patient engagement.

If we wish to create a scalable and equitable healthcare system to manage the increasing care needs of our expanding and aging population, we must expand our notions of healthcare beyond visit-based care; we must adopt additional communication technologies and optimize the selection based on the clinical situation being discussed, the availability of technologies, and the preference of those communicating.

The third element that relates to communication is a shift in mindset, both on the part of the patient and the physician, to make healthcare more effective. Patients must take responsibility for their health, and must recognize that healthcare is not a spectator sport, it is a participatory sport. They should participate in decision-making with their physicians, ask questions, and be willing to invest time and effort in understanding their conditions. Physicians, for their part, must be willing to share information with their patients (including access to their records), and must respect that although they may be experts in their area of healthcare, their patients will always be experts in themselves. Moreover, they must encourage their patients to learn more about their disease and engage them in the development of shared care plans. They should not fear but should encourage patients as they evolve into e-patients: those that are engaged, empowered, equipped, enabled, and educated.

In this book Nancy Finn, a writer, editor, and consultant with a strong background in digital communication technology explains why it is so important to be an e-Patient and discusses the tools that facilitate patient engagement. She brings relevance to provider tools such as e-prescribing, electronic medical records, and health information exchanges, and even reviews advanced technologies such as telemedicine, home monitoring, robotics, precision medicine and artificial intelligence in health. Along the way she discusses our often-dysfunctional healthcare system and how e-Patients can leverage technology to overcome deficiencies in the system and improve their care.

Technology is important and interesting, but is not an end unto itself. It is an enabler of new models of care and can lower barriers to patient engagement. That is why this book is so important to modern patients and caregivers. It is my sincere hope, that through books like this, every patient will become an e-Patient.

Daniel Z. Sands, MD, MPH, is an Assistant Clinical Professor of Medicine, at Harvard Medical School, and a staff physician at Beth Israel Deaconess Medical Center, Co-Founder and Chair Emeritus, of the Society for Participatory Medicine.

INTRODUCTION

The recent pandemic has awakened all of us to the need for every patient to become empowered, engaged, and educated. The fact that more than 1 million individuals, in the United States alone, died from the Coronavirus, points to the need to protect ourselves, by thinking about our own health and the health of others.

COVID-19 made us aware of the fragilities of our healthcare system. We now realize how the reckless disregard for our most vulnerable citizens impacts all of us, and reinforces the need for equity. We were confronted with the dangers of not having adequate control over our supply chain that provides critical equipment and materials that include: ventilators, test kits, respirators, gloves, masks, face shields, hand sanitizers, critical medications, and supplies needed to treat us when we are seriously ill. As empowered patients we must take steps to see that our leaders mitigate supply-chain issues in the future, and put in place fast response systems, so that we will be safe.

Digital technology and digital apps exploded during COVID and enabled patients to become more actively engaged in handling their own health. New applications, such as zooming with healthcare professionals, our caretakers, family and friends, and remote monitoring of our vitals from our homes, now enable us to share data with our clinicians. The availability of our digital health records, and patient portals helped us easily interact with health providers and doctors. Digital health records have made it possible for an accurate summary of our health history, conditions, medications, and test results to be available to us and to our clinicians, anywhere, anytime. Patient portals, which are secure online websites that house our data in one central location that can be accessed with an internet connection, ensured, that despite the pandemic, patients experience continuous, coordinated care. When the pandemic made it impossible for many of us to see our physician in a live setting, these portals enabled us to have remote visits, and truly feel confident we were being taken care of with full information at the point of care.

Telemedicine, which involves using fiber optic or broadband technology to support clinical care when distance separates patients and providers, was vastly expanded, with the passage of the Coronavirus Aid, Relief, and Economic Security Act, (CARES) that includes provisions to broaden coverage of, and provide grants to support the greater use of telehealth services, paid for by Medicare, private insurance, and through other federally funded programs. The Act also includes several changes to the Medicare program, that eliminate certain requirements, including: face-to-face encounters and delays that resulted in scheduled payment reductions. CARES also increased Medicare payments for the treatment of patients with COVID-19, permitting a ninety-day supply of prescription drugs during the COVID-19 emergency, as well as coverage for any COVID-19 vaccine without cost-sharing by patients.

Mobile devices including cell phones, smartphones, and tablets enabled patients, doctors, and researchers to access and monitor health issues and increase our understanding about physical and mental health and well-being. Mobile health support was proven to be simple, but effective. For example, anyone with the ability to send and receive text messages can be contacted when necessary. There are also new sensor apps built into mobile devices that are able to collect vital health information and, if the app detects a sudden change, or emergency, it will send a signal to the nearest EMT using the built-in mobile phone GPS to identify your location.

Remote Patient Monitoring (RPM) is a broad term that refers to the use of a variety of medical devices, used at home by patients to monitor and manage chronic conditions. These devices include wearable clothing and jewelry, home health monitors, and online apps. RPM technology electronically tracks and transmits real-time information from patients to clinicians, seamlessly, on many platforms, collecting information such as weight, blood pressure, and heart rate using an external cuff, a special scale, or a camera connected to an iPhone, iPad, or laptop. RPM improved technically and became more affordable during the pandemic years. A study by Insider Intelligence projects that by 2024, RPM tools will be used by more than 30 million patients.

During the pandemic, social networking became one of the standard ways that people remained in touch, to talk about feelings and exchange opinions. People accessed social networks to connect with friends, family, business associates, and other patients, while remaining socially distanced. PEW Research Services, a nonpartisan fact tank that informs the public about the issues, attitudes and trends shaping the world, published a survey in February, 2021 that interviewed 1,500 adults in America. The survey found that nearly 100% of the participants use social networks; over 50% are registered with several; and many people use social networks to connect and learn about health issues.

This book explains what it is like to be an e-Patient who deploys digital technology that we all use every day: email, the internet, and mobile devices, to communicate with healthcare providers and ensure the information you need is available at the point of care. The book provides details about digital technology and innovative communication tools that ensure that your personal health information is available, no matter where or when you might need care. The book identifies and outlines the benefits and barriers to adoption of new technology that has the potential to transform your care, as well as ensure that the care you seek is responsive and right for you; that it is easily and efficiently delivered by your providers in a trusting patient/clinician relationship.

There are significant barriers to adopting innovative technologies, including: cost; privacy issues; willingness of patients to use new devices; and misunderstanding among various stakeholders, including: technology developers, entrepreneurs, healthcare executives, investors, doctors, and the patients. This book addresses these barriers.

E-Patients Live Longer is filled with suggestions, resources, and thought-provoking anecdotal stories that explain why health information technology and participatory medicine is so important. Based on hundreds of interviews with individuals who are on the cutting-edge of e-health and medical practice, this book concludes with a discussion of exciting developments such as genetics, robotics, precision medicine, and other as yet unknown discoveries to make healthcare better for everyone. Once you have completed this book, you will become an e-Patient, who understands digital technology and new, innovative communication tools. As the pandemic demonstrated and this book will prove, e-Patients who understand what is happening in this digital world that is rapidly evolving will help themselves live longer and enjoy a better quality of life.

Content Disclaimer: The anecdotal stories sprinkled throughout the book are composites of real patient stories that were relayed to the author by healthcare professionals or are from published stories and used with permission as noted in the endnotes. The names have been changed to protect patient privacy.

Medical Disclaimer: This book is not a book about specific diseases, or how to find the right cures or treatments. It is a book about how to use digital communication tools to empower yourself as a patient for better, safer healthcare.

Because of the dynamic nature of the Internet, any web addresses or links contained in this book may have changed since publication and may no longer be valid. The views expressed in this work are solely those of the author and do not necessarily reflect the views of the publisher, and the publisher hereby disclaims any responsibility for them.

TABLE OF CONTENTS

Chapter One

Power Up and Communicate
with Your Healthcare Providers

Be what you are and say what you feel because those who
mind don't matter, and those who matter, don't mind.

Bennett Cerf, Shake Well Before Using: A New Collection of
Impressions and Anecdotes Mostly Humorous (1948) p. 249.

The Patient Experience

*Jim has asthma. At his office visit, he tells Dr. Oakes, his primary care physician (PCP)
of fifteen years, that despite faithfully taking medications prescribed, he is wheezing a lot
at night and sometimes awakens short of breath. Dr. Oakes reviews Jim's electrocardiogram
(EKG), finds it to be normal, and listens carefully to Jim's heart and lungs. Everything
seems to be fine. While Jim waits, Dr. Oakes searches Jim's records to locate his list of medications. They talk for a few minutes about what could be causing this problem. Dr. Oakes
hands Jim a new prescription, which Jim will take to the pharmacy to fill. Dr. Oakes also
writes orders for a chest X-ray and a pulmonary function test and suggests that Jim return
in a month to review the results of these tests and his blood work. All of this paperwork
takes time away from the all-too-short visit, which lasts a total of fifteen minutes.*

This is a typical encounter between a physician and a patient. With time
restrictions on doctors imposed by insufficient primary care reimbursement

and a paper record system that takes the doctor several minutes to search, this office visit is structured so that there is barely time for Dr. Oakes to properly examine Jim. There is never enough time to formulate an evidence-based diagnosis and discuss a treatment plan, let alone talk to Jim about what else is on his mind. Dr. Oakes also writes his comments as they talk, so he is not able to give his full attention to what Jim is saying. The experience is frustrating for both Dr. Oakes and for Jim.

Sandra, a retired teacher with hypertension, has been a patient of Dr. Clarke for twenty years. When Sandra arrives for her annual visit, she is given a clipboard with a paper form that asks her to list all of her medications and fill in her medical history. She observes that this is the same form she filled out last year, as she diligently tries to remember all of the details required. Sandra is escorted into the exam room by a nurse who asks about her general health; checks Sandra's vitals; gets her weight; draws blood; and does an EKG—all part of the routine annual checkup. When Dr. Clarke comes in, she greets Sandra, quickly glances at her update form, and reviews a piece of paper that Sandra has brought with her that has a list of blood pressure results Sandra has tracked over the past several weeks. They talk briefly about headaches that Sandra has been experiencing. After a quick but thorough examination, Dr. Clarke asks Sandra to get dressed and meet her in her office, where she sits at a computer terminal reviewing Sandra's electronic health record. With her eyes on the screen, Dr. Clarke tells Sandra that the exam was fine. She asks Sandra to continue tracking her blood pressure and send a weekly email to the nurse, who will review the data for anything unusual. She enters a prescription for the headaches into the computer, which automatically checks the new medication for interactions with other medicines that Sandra is taking and sends the information electronically to Sandra's pharmacy. Dr. Clarke suggests that Sandra schedule another appointment in six months or sooner if the headaches persist or if there are any other problems. She encourages Sandra to email her if she has further specific questions about the treatment plan.

This office visit, which also lasts approximately fifteen minutes, is a little closer to what a twenty-first-century e-Patient visit with the doctor looks like. Sandra's information is in an electronic health record, and her doctor uses e-prescribing to send Sandra's medication to the pharmacy, reducing the possibility of medical errors from illegible handwritten prescriptions and insuring a more comprehensive check of drug interactions. The fact that Sandra is monitoring her blood pressure empowers Sandra to be more involved in her healthcare. Hopefully, this will keep her stable and out of the emergency room. Although the visit is pleasant and efficient, it does not foster the kind of communication and discussion that physicians and patients should routinely experience. The doctor is rushed. She is facing the computer and has her back to Sandra while they talk, so there is little eye contact and personal dialogue. The use of digital technology is a step in the right direction in so many ways, however, the technology can be intrusive. Although Sandra has a way to connect with her physician between visits, which helps her adhere more rigorously to her physician's recommendations, there are things about their interaction that are unsettling.

What Do You Want from Your Visit with Your Doctor?

When you go to the doctor, you are seeking solutions to specific health issues. You want to be able to talk with your doctor, to feel welcomed, and to be assured that your issues are going to be addressed with respect, competence, and confidentiality. Here are suggestions regarding what to look for:

- A doctor who gives you his or her full attention and listens to what you have to say.
- A doctor who is thorough in examining you and checking your medications, blood pressure, heart, lungs, etc.
- Someone who is caring and compassionate and with whom you have a rapport that enables you to talk freely about your concerns.
- A doctor who is committed to helping you solve your health issues in the most efficient and effective way possible by offering communication channels beyond the limitations of the office visit.
- A physician who outlines all of your options for treatment and directs you to information resources on the internet or articles that give you more detail than can be covered during the office visit.
- A physician who provides you with specific well-explained guidelines to follow when you leave the office.

What Makes You a Good Patient?

Doctors are human, healthcare is complex, and time spent with each patient is restricted. As an e-Patient, you must assume an appropriate level of responsibility for your care. You should come to the office visit prepared to provide a complete list of medications that you are currently taking, as well as up-to-date information on other providers you have seen and issues you have been dealing with. You must ask for explanations and guidance when your doctor's diagnosis or orders are confusing or contradictory. You must follow through with those orders, including taking and finishing all medications in the manner they are prescribed and complying with your doctor's request to keep records of your blood pressure or blood sugar. It is important to be forthright, even if it involves embarrassing personal issues. You should come to your annual checkup equipped with a pen and paper to write down comments that are difficult to remember during the visit. Some of the questions you might want to ask and answers you should record include:

- What do you think is causing my symptoms?
- What does this mean?
- Do I need tests?
- What kind of tests?
- When and how will I receive the test results?
- What are my treatment options?
- Do I want to see a specialist and why?
- Do I need medication?
- What kind of medication?
- What are the side effects of the medication?
- Are there lifestyles changes that I need to consider?
- When should I see you again? Can I ask you questions via email?
- Can I view my test results?

Effective doctor-patient communication is critical to the delivery of appropriate healthcare. Your relationship with your medical provider and your interactions during the clinical encounter can influence your medical outcome and your satisfaction with your care. Open dialogue is based upon your doctor taking the time to listen to you, and your providing honest and complete information. There is no holding back information on drugs that you are taking, both prescription and over the counter. Your physicians need full, factual information about you in order to make a proper diagnosis. With both Jim and Sandra, the rushed encounter left little time for either patient to chat about their current life situation, their stress, their feelings, or other concerns that could have a serious impact on their health.

Studies of asthma patients reveal that when it comes to taking the oral steroids that are often essential for controlling asthma, many patients do not follow the doctor's directions for taking the medication. This is often due to the fact that these patients did not have the opportunity to discuss fully with their physician the reasons why they need medication, what is an appropriate dosage, and when and how to take the medication. (Inhalers can be very different from one to another in the way they deliver a dose of medicine). As a result, the patients ignore the doctor's orders. Although they typically fill the prescription, they take their dosage erratically and often stop the medication completely after a few weeks or months. Eventually they end up back at the doctor's office or in the emergency room. Simple communication could help resolve this problem of noncompliance and keep patients more stable and able to control the disease. (1)

According to the Centers for Disease Control and Prevention (CDC), doctor visits in the United States have increased annually, although face-to-face visits dropped significantly during the pandemic, replaced with telehealth visits. Each of these visits requires an intricate patchwork of information processing and communication. Most physicians have transitioned to electronic medical records), over 90%. Among physicians who do store their records in computer systems, most cannot exchange patient data electronically with other providers that you might see, which can result in an information gap. As a result, your paperwork sits in isolated files in the possession of a myriad of providers, including your primary care physician, your specialists, hospitals where you have been admitted, labs where your blood and other tests are sent, physical and occupational therapists, alternative medicine specialists, your pharmacy, your dentist, and your health insurer.

In a study conducted by the Agency for Healthcare Research and Quality (AHRQ), which is a part of the Department of Health and Human Services, clinicians reported that in a representative sample of over 1,500 clinical visits, important information was missing in 13.6% of the cases. Among these visits, the most common missing information was lab and radiology results (45 percent and 28.2%, respectively), letters or dictation containing clinical information (39.5%), and patient history or physical exam findings (26.8%). When your physician does not have all of the facts, serious consequences can occur, including the need for expensive repeat tests or the possibility of medical error. (2)

Time, Tools, Teamwork

Talking to your doctor requires time, tools, and teamwork. You typically spend more of your day communicating (using the telephone, written word, and computer, or talking with other people) than in any other single activity. These habits provide a unique set of well-developed communication and listening skills. When it comes to your healthcare, those skills matter. More than 80% of the information that your doctor needs to make a correct diagnosis comes from what you say and your medical history. The rest comes from the examination and test results. Unless you accurately describe your problem, it is difficult for the doctor to determine what is wrong. Greater use of digital communication tools—electronic health records, email, electronic databases, patient portals, and e-visits could make a vital difference in assisting your physician in making a more accurate diagnosis and prescribing a treatment, enabling you and your physician to establish a better dialogue. Sending a medical history update form to your doctor in advance of your visit enables your physician to have the required information about you. Communicating with your providers more frequently between visits, for non-urgent issues, frees up time for you and your physician to address critical issues during the clinical visit.

Diana is a fifty-seven-year-old consultant who suffers from chronic arthritis and related autoimmune problems. Ten days before her scheduled visit, Diana receives an email from her PCP's office with a link to the portal that Dr. Cooper shares with her patients. The email reminds Diana to go to her private section of the portal and enter any recent problems, test results, or meetings with medical specialists or therapists she has had since her last visit. Early on the day of her appointment, an automated computer program sends a checklist of all the patients on her schedule to Dr. Cooper's smartphone. Just before Diana's visit, Dr. Cooper is able to quickly glance at Diana's portal entries and determine the agenda for their visit. When Diana arrives at the office, she goes directly to a computer terminal, where she scans her medical card. This notifies the staff that she has arrived and brings up her electronic health record. She is escorted to an examining room, where a nurse takes her vitals and keys that information directly into her the record. She leaves that screen visible for Dr. Cooper, who is able to look very quickly at the computer screen to assess Diana's condition. As a result, Dr. Cooper can focus on Diana without the distraction of having to look up information. Dr. Cooper changes a couple of Diana's medications and sends a prescription directly from her computer to the pharmacy. She also sends an email to Diana's rheumatologist to update him and posts links to Diana's patient portal site with her next appointment, all in a matter of seconds. Dr. Cooper reminds Diana that if she has any questions, she can send her an email. On her way out, Diana goes to a kiosk where a short survey is on the screen that asks a few questions about her satisfaction with the visit.

The Empowered Patient

Becoming an e-Patient requires you to change your philosophy regarding your healthcare. Instead of viewing the doctor as the guardian of your care, you must become an integral part of the medical team taking care of you. There are several actions that will empower you:

- Do not stay with the same doctor if he or she is not meeting your needs. Choose a doctor who uses best practices in diagnosing and treating your problems.
- Gather a healthcare team with whom you can collaborate to address your health concerns. Be sure there is two-way discussion about your healthcare options.
- Assume responsibility for collecting your health history, test results, and provider notes, and bring that data with you when you meet with your physicians.
- Always arrive at your healthcare provider's office with a list of issues prioritized according to what is most important to you.
- Elevate the quality and efficiency of your interactions with your healthcare providers by setting clear expectations regarding what you expect for a visit with your physician.

- Insist on fast access to medical attention when you have a serious problem, including same-day visits, whether online or in person.
- Seek advice and information on appropriate internet resources to help you understand your problems, issues, medication side effects, and treatment options.
- Understand your health insurance choices.
- Arrange for a patient advocate when you are too ill to be your own advocate.
- Speak up when you are uncomfortable with what your physician might or might not be doing.

Talking to your doctors and getting what you need requires you to participate fully and take personal actions. You also need to protect yourself from shoddy care, medical error and overpayment. Here are a few questions to ask:

- How experienced is my doctor?
- Does he or she satisfy my needs?
- Do I need to see a specialist?
- What hospital best fits my needs? Is it the big academic medical center forty miles away or the community hospital that is walking distance from my home?
- Is there a difference in what I would pay for a procedure or operation at these institutions?
- Does the difference align with the level of quality I want to experience?
- Does my doctor provide me with options for treatment when appropriate?
- Does my doctor explain the details of my medications and help me understand side effects and the schedule for taking the medication?
- Does my doctor help me find worthwhile web resources?

Key Points

1. In the fifteen-minute office visit, there is not enough time to discuss the complex issues that most patients bring with them. Alternative communications are necessary to provide your doctor with information in advance of the appointment, or to enable you to talk with

your doctor or nurse after the appointment to clear up unanswered questions and issues that arise. Options include planned Q&A online sessions with your doctor following the visit and email queries sent to a qualified healthcare professional who will answer your concerns.

2. Good communication includes honest, complete information that is free of jargon and involves good listening skills by both doctors and patients. Consider what might have changed in your life over the past several months to warrant communicating with your provider. Think of ways to improve communication between yourself and your providers so you are better informed and they are current with your individual health issues. Don't wait for an emergency.

3. With increasing patient loads and decreasing numbers of healthcare providers to meet demand, digital communication technology must fill the gap to make healthcare delivery more efficient, cost effective, and higher quality. This includes electronic health records, email, online resources, and e-visits.

4. The e-Patient is an individual who has thought about healthcare and has a plan that includes evaluating the effectiveness of your healthcare team. As an e-Patient, you need to be aware of your information needs. Take responsibility for your education about health matters and initiate a productive Q&A with your physician.

5. The e-Patient becomes an integral part of the team by setting goals, evaluating providers, ensuring that ongoing communication is a priority and taking the lead when issues or questions arise.

Notes

1. Helen K. Reddle, et al., on behalf of the American Thoracic Society/European Respiratory Society Task Force on Asthma Control and Exacerbations, "Asthma Control and Exacerbations: Standardizing Endpoints for Clinical Asthma Trials and Clinical Practice," American Journal of Respiratory and Critical Care Medicine, 180 (July 2009): 59–99.

2. Wilson Pace, MD, et al., "Missing Clinical Information During Primary Care Visits," Journal of the America Medical Association, 293(5) (February 2, 2005):565–71.

CHAPTER TWO
Digital Health Records Could Save Your Life

> Information is the lifeblood of modern medicine. Health Information technology is destined to be its circulatory system. Without that system neither individual physicians nor healthcare institutions can perform at their best or deliver the highest quality care.
>
> David Blumenthal, the National Coordinator of Health Information at the Department of Health and Human Services
> New England Journal of Medicine, January 21, 2010.

In the fall of 2005, Hurricane Katrina destroyed the health records of thousands of individuals in Louisiana and Mississippi. Medical officials had been warning of catastrophic medical events following a natural disaster, terrorist attack, or uncontrollable outbreak of disease. The disappearance of thousands of medical records meant that healthcare providers working with these patients under emergency conditions did not have the required information at the point of care. The paper records that their physicians had kept were destroyed in the flood, and the few medical institutions that had electronic records did not back them up to a remote storage facility. With the available technology at that time, this situation should never have happened. Hopefully it will never happen again.

EHRs and PHRs

For centuries, handwritten notes made up a patient's health record and provided the basis for a treatment plan. You did not have access to this information beyond the verbal directions relayed to you by your physician. Today, your health information is kept in an electronic health record (EHR), which includes information that you provide to your physicians when you fill out those paper forms you are handed when you arrive at the doctor's office. Your physicians also record data, based on discussions, observations, lab procedures, medications you are taking, and other information relevant to your care such as X-rays, scans, and hospitalizations. In other words, your electronic record is a complete profile of your current and past health. Many EHRs also integrate scheduling and billing functions and include decision support tools for your doctor, such as drug alerts, allergy alarms, and other treatment options that enable your doctor to care for you more effectively.

In addition to an electronic record, some of you might also have a personal health record (PHR), a computerized record that you personally create that identifies all of the providers you see, the medications you take (prescription and over-the-counter), procedures that you undergo, and your medical history. If all of the patients impacted by the Katrina floods had personal health records, their data might have been easier to retrieve when they needed to seek care following the hurricane. You provide all of the information that is stored in your PHR and are tasked with continually updating it. The information in both the EHR and the PHR are technically owned by you. Your EHR is generally held and maintained by your provider on your behalf, while your PHR is generated and maintained in a system that you have chosen.

Electronic health records provide your doctor with instant information, including notes from prior visits, lab tests, comments about treatments and medications, and consults with specialists. Your physician can also see the results of labs and procedures that were done years ago and pull up a history of your blood pressure or blood sugar readings, procedures, and prior tests. With an EHR, if your doctor recommends that you see a specialist, a referral can be generated right from the digital health record system to set up your appointment. Your physician may put a reminder in your EHR to follow up with the results of a test that has been ordered, or vaccines or treatments that need to be done in the future. If your doctor or a specialist you are seeing practices at a hospital that has converted to a compatible EHR system, your emergency room visits, consultations, lab test results, and inpatient records will also be in your EHR and available to all of your providers. This will make your life easier because you will not have to struggle to remember your medical history, including all current and past conditions, treatments, and medications, when you see a new physician. It will also make your life safer because there is less duplication of tests and procedures and less room for human error when all of the information is right there. If the doctor is ordering a new medication for you, those prescriptions can be sent instantly from the EHR to the pharmacy, and the system will check for adverse drug interactions with what you already take before you leave the doctor's office. If the system indicates that the medication could cause a problem for you, the doctor can order something else while you are right there.

As important as it is for your physician to have your electronic health record available during an office visit, it can be critical that your EHR is available when you go to an emergency room (ER). Imagine an eighty-two-year-old person coming to the ER with a stroke. A clot-buster drug, if administered in the first few hours, can save that person's life; however, the clot-buster is dangerous if that person is on a blood thinner, has had recent surgery, or has a history of bleeding. With a comprehensive EHR, the doctors treating that individual will know all the details that could fatally impact the treatment of the immediate critical symptoms, including the individual's health history, medications and dosages, allergies, or recent problems. The lack of this information could be catastrophic.

The instant access to information enabled by an EHR results in a better information exchange between you and your doctors, as well as between your doctor and other providers who might consult on your case. All of this should improve care coordination, reduce medication errors, speed up referrals, and empower you to be more involved in your own care. By consolidating all of your health data in one place that is accessible by all of your providers and by you, the architects of the meaningful use statutes assume that these EHRs will foster better communication, engaging you and your family in managing your health.

It is a fact that your healthcare provider today must cope with the enormous communication requirements that have markedly increased in the twenty-first century. The spread of digital records, email, and the internet has inundated you and your doctors with information. In the average annual visit to the doctor, numerous tests are done as a routine matter. Those tests can generate several pieces of information coming back for analysis. The analysis often leads to more tests, resulting in a lot of information to be interpreted and integrated. Years ago, the average patient could be taking two medications that had to be tracked and followed. Today many of you take multiple medications, and all need to be tracked. Neither you nor your doctor can possibly keep in your head the results of your tests, the dosages of your medications, the side effects you may experience, and the alternatives that best fit your patient profile. As a patient, you expect complete, instantaneous information, including what is wrong with you, what to do about it, who to talk to, and why it is important to thoroughly filter all of this information. There is also evidence that electronic health records include data that enables healthcare providers to better care for underserved populations and thus reduce healthcare disparities. Additionally, having information electronically provides the tools that should enable healthcare providers to analyze and take action on public health issues that hopefully improve healthcare for everyone.

Patient Benefits

You benefit from electronic health records for the following reasons:

- They generate reminders for screening tests that might otherwise be forgotten.
- Information can be shared with all of your providers and with family members who are concerned with your care.
- They provide necessary information for major surgery and procedures such as preregistration information, specific details about the procedure including how to prepare and what to expect, advice on home care, and other necessary follow-up protocols.
- They include instructions for acute and chronic conditions, such as instructions in self-monitoring, lifestyle, medications management, action plans, what to do at home, what signs of problems or improvements to look for, and when to call if symptoms develop or improvements don't occur as expected.
- They provide you with suggestions about how to prepare for scheduled visits, including questions to ask your provider and biometric instructions (e.g., the need to fast before a test).

Personal Health Records (PHR)

Jane, age fifty-three, lives in a rural community in South Dakota. She has been diagnosed with breast cancer, which was spotted on her annual mammogram. Her primary care physician refers her to a cancer center two hours away from her home for further screening. The doctors there decide that Jane needs surgery, followed by chemotherapy at the cancer center one hundred miles away. Prior to visiting the cancer center, Jane is instructed to collect all of her medical records, including those held by her primary care physician and gynecologist, and the X-ray films filed at her local hospital. While dealing with the trauma of a cancer diagnosis, Jane also has to navigate through the bureaucratic system to locate and copy records.

With the assistance of her PCP, Jane decides to set up a PHR that includes an extensive outline of her family's history of breast cancer, her own medical history, a list of her medications, results of labs and tests, her mammogram reports, and other scanned images. She shares the PHR with all of her doctors from the cancer center, who also provide her with the surgery and pathology reports and specifics on her treatment. Thus, both her local doctors and the physicians from the cancer center are able to follow her progress and have full information on her treatment. Jane also backs up a copy of her PHR on a CD that she carries with her. At one point in her treatment, she develops a high fever and has to go to the emergency room at the local hospital. By accessing her PHR, the doctors there are able to avoid a potentially serious adverse medication reaction. Jane is an e-Patient who uses her PHR to communicate and ensure her well-being.

Options for Creating a PHR

It is no exaggeration to say that a personal health record could save your life, particularly when a comprehensive electronic health record is not available. There are many ways to develop a PHR, and the method you choose is partially dependent upon what is available to you from your healthcare providers or your health plan. With all of these options, except paper, you would need to have your login information with you. The following are suggestions for how to implement a personal health record.

1. You can resort to the old-fashioned handwritten synopsis of your medical heath and history on paper. This gives you a baseline from which to work and is good for an emergency—if you remember to bring it with you. This format is not useful in a shared environment or in an emergency if the paper PHR is at home.
2. If the medical institution that you go to has a patient portal populated by data from your clinicians' electronic health record, you may be able to add your PHR where it could be saved to the portal's database. In this model, you will have some access to your provider's electronic record data, which is connected to the existing clinical information systems. The electronic health record combined with your PHR represents a comprehensive synopsis of your medical information.
3. Your health insurer (e.g., Aetna, Blue Cross Blue Shield, or United) may provide you with the opportunity to create a PHR that will be

populated with administrative claims data, such as discharge diagnoses, reimbursed medications, and lab tests ordered. The limitation of this PHR is that it is not a complete picture of your health profile, and the data could be inaccurate.

4. Many companies offer their employees personal health records that are also based upon claims data and benefit information. Typically, these PHRs are maintained by an independent outsourcing partner and it is your responsibility to continually update the information.

5. You could opt to put all of your personal medical information on a USB memory stick attached to a key chain. When you want to access your health information, you plug the chip into a standard USB slot in any computer, including the systems in your healthcare provider's office or the computers in the hospital emergency room. The memory stick will download the information in your PHR into the computer where a doctor anywhere in the world can view it on the screen, print it out, or transmit the information to another provider. This system does not depend upon a doctor having your electronic health record. The information on the memory stick can be encrypted so that it can only be accessed with a password that you provide, thus protecting your data from unauthorized viewing.

6. There is also the PHR on a smart card, which is similar to a social security card or an ATM card. The PHR smart card includes a patient's complete health history and information on individual illnesses, as well as demographic and geographic information, insurance data, medication lists, healthcare providers' names and contact information, and insurance records. You would carry your PHR smart card at all times, just like a credit card. Your primary care physician would have complete access to the medical information on the card. When you visit your PCP, the card is automatically updated with new information. Smart cards use digital signature technology with encryption and are thus more secure and private than many other forms of PHRs. In case of an emergency, the information on the smart card can be retrieved to access your medical information. However, a smart card can only be read on a compatible smart card reader, which is not available in most doctor's offices or emergency rooms. The smart card solution works best

within a specific geography in a collaborative health organization or a group practice where card readers are available.

7. If you have stored your PHR in any of the online hosting sites, you can ensure that the information is always available by having a class II implantable radio-frequency-transponder system implanted in your arm or leg. Although still in the experimental phase but approved by the FDA, this implantable device, which is the size of a grain of rice, can be activated by a handheld scanner. The radio frequency identification (RFID) chip emits radio waves read by the scanner as a unique, sixteen-digit patient identification number. Your healthcare provider would then use that number to access a computer database containing the medical information you previously entered and regularly update. This information includes allergies, your medication list, your medical history, the name of your primary care physician, etc. This approach presupposes you are wearing a bracelet or have some other means of identifying the implanted chip and that the facility caring for you can find the information on the internet and use it as needed.

You should include the following information in your personal health record:

1. All of the medicines you are taking, including prescription and over-the-counter medicines, dietary and herbal supplements, vitamins, and minerals. For each medicine, include the name of the doctor who prescribed it, why you are taking it, how much you take, and any special instructions.

2. A list of your allergies, including drug or food allergies.

3. Your medical history, including major health problems and chronic conditions such as asthma, hypertension, or diabetes; hospitalizations and the reasons for them; surgeries and broken bones; mental health issues including drug and alcohol issues, depression, or bipolar disorders; and illnesses such as pneumonia even if you were treated without hospitalization. Also include major health problems in your immediate family, such as heart disease, stroke, cancer, or diabetes.

4. Information that is needed in an emergency, such as whether you have a pacemaker, implanted device, or hearing or vision problems.

5. General information including identification such as a driver's license, who to call in an emergency, the name and phone number of your primary care physician, your insurance card, an organ donor card if you have one, a copy of your healthcare proxy, a living will and power of attorney, your pharmacy name and phone number, and any records of insurance claims and payments.

Accurate and timely health information is a crucial element in the decision-making process during a medical encounter and essential in providing you with continuity of care. Inadequate or misleading patient information can lead to medical errors, inaccurate assumptions, and increased cost. Providing your physicians with access to every detail of your medical history is difficult. Striking the balance between adequate and too much information is tricky. Your electronic health record and your personal health record have emerged as ways to address these issues and maintain that balance.

Case Studies

The following examples of how electronic records are used in key institutions illustrate why they are a critical component in providing you with safer, better quality continuous healthcare.

Veterans Administration Healthcare System

One of the earliest adopters of electronic health records is the Veterans Administration (VA), that, in the mid-1990s, embarked on an ambitious effort to improve the coordination of care by implementing an electronic health record, known as the Computerized Patient Record System or (CPRS). CPRS provides a single interface through which VA providers can update a patient's medical history, submit orders, review test results, review drug prescriptions, and perform other functions to support clinical care delivery and the promotion of wellness. The system has been implemented at all VA medical centers nationwide, and at VA outpatient clinics, nursing homes, and other sites of care. It allows any authorized person to look at patients' records, including everything from a nurse's note written during a hospital stay to the result of a blood test drawn at a clinic visit to the film of a coronary angiogram done in a cardiology lab.

The VA system also enables health information exchange among the more than 1,400 VA facilities, so that the medical record with the patient's history is available online to all doctors in the system, no matter where that patient shows up for treatment. Repeat MRIs, scans, blood work, or other unnecessary tests are avoided, and there is much safer medication management for the patient.

Kaiser Permanente

Kaiser Permanente (KP) is the largest private not-for-profit integrated healthcare delivery system in the United States. The system includes Kaiser Foundation Hospitals and their subsidiaries and the Permanente Medical Groups. KP uses technology-enabled care and has developed many tools that encourage its patients to communicate with their providers and manage their health. KP providers were among the first to adopt a full-featured electronic health record.

Health Information Exchange

In order to maximize the benefit of an EHR, it needs to be part of a total health solution that includes health information exchange (HIE). HIE provides the capability to electronically move clinical information among disparate information systems, such as different hospitals, while maintaining the authenticity of the information being exchanged. The benefits of sharing health information among patients, physicians, and other authorized participants in the healthcare delivery chain are obvious. You often see a variety of providers to meet your health needs. For example, for treatment of a chronic condition, you could visit your primary care doctor, a specialist, a clinical laboratory, an imaging center, and a pharmacy. Each of these providers may maintain records of medical treatment, laboratory results, medications, a health history, and personal information about you that may also be needed by other providers. This information may only be available in printed format. As a result, access to your specific health data could be missing at the point of care and delay treatment.

If your cardiologist is able to obtain an abnormal laboratory result electronically, this faster access could enable that doctor to perform earlier intervention for a potentially life-threatening condition.

There are also built-in safeguards with health information exchange to keep people from abusing the system. One hospital found that information obtained through its health information exchange helped its emergency department physicians ascertain that a patient who was requesting medication for pain had been in five area hospitals in seven nights seeking pain medication. Needless to say, the system allowed the professionals to stop the abuse.

In a pilot program that promises exciting opportunities for health information exchange in the future, KP and the VA have partnered to exchange health information across a Nationwide Health Information Network (NHIN). Use of this network enables clinicians from the VA and Kaiser Permanente to obtain a more comprehensive view of a patient's health across both systems. Unfortunately, the example of these two institutions using health information exchange is not repeated with other institutions throughout the country.

The Boston area is, perhaps, the only location in the country where there are five renowned academic teaching hospitals: New England Medical Center, Boston Medical Center, Beth Israel Lahey Health, the Massachusetts General Hospital (MGH), and the Brigham and Women's Hospital, three of which are affiliated with Harvard Medical School. Additionally, there are several specialty hospitals in Boston: Dana Farber Cancer Institute, Joslin Diabetes Clinic, Boston Children's Hospital, Shriners Hospital For Children, Boston There is no health information exchange among these hospitals. This means that a patient who is in a serious accident in the Boston area and who is transported unconscious to the nearest hospital could be taken to any of these institutions, and there is no way for their health information to be made available for the doctors who are caring for them in the ER. These hospitals, with the exception of Massachusetts General Hospital (MGH) and the Brigham, which are both under the Partners' healthcare umbrella and therefore use the same electronic health record, cannot share records of patients. The others have different health systems and different electronic records and their information cannot be accessed by the doctors in the emergency department for prompt safe care where health conditions, allergies to medicines, and other key factors are available to the doctors caring for them. The technology is available to make this happen, but the desire among the hospital systems is simply not, when it should be a given.

Key Points

1. Make healthcare a priority in your life.
2. Talk with your providers and find out if they have an electronic health record for you. If yes, ask how you could gain access to it in an emergency. If no, ask when they envision having the technology to share your record with you and all of your providers using secure communications.
3. If you see specialists, ask your primary care physician whether or not their systems are interconnected to the specialist so your records can be shared.
4. Whether or not your physician has an electronic health record, make it a priority to create a personal health record that includes all of your medications, previous procedures, surgeries, chronic conditions, allergies, family history, and other relevant data. This PHR must contain information from your entire health history, and you must keep it up to date.
5. Be sure to monitor who has access to your PHR. Keep it in a secure environment, on your person, or in a secure site on the web.
6. Find out if there is a regional collaborative that your primary healthcare group is affiliated with. Inquire whether or not you can go to any provider in the collaborative and have your information shared.
7. Become an advocate for better communication, and the installation of a health information exchange system, at the least on a regional level, so that information sharing that could save lives is available.

Table 1

What to Put into Your Personal Health Record?

Health History	Illnesses, injuries, surgeries, chronic conditions hospitalizations, broken bones, mental health issues; include all dates, duration, and treatment. Family related health issues.
Emergency Contact information	Healthcare proxy, names, phone numbers, alternate contacts
Health Insurance	Policy number, type of insurance, duration of insurance contract, co-payment information
Medications	Prescriptions including name of pharmacist; all over-the-counter medications, dietary & herbal supplements, vitamins and minerals, dosage and frequency
Allergies	Medications, foods, airborne allergies such as dust and pollen and mold, fabrics,
All of your physicians – primary care and specialists	Name, address, phone numbers, hospital affiliations
Special situations	Permanent implants, such as joint replacements, pacemakers, special hearing or vision problems
Personal information	Living will, driver's license, employment history, employer, (Name, address, phone), names of family members, (spouse, children, parents)

CHAPTER THREE
Continuous Care: Email, Portals, and Smartphones

> Information technology has freed information from the constraints of geography and time. Information in electronic form can be made available at any place, at any time, and is not tied to a physical location. Further, software tools can make communication detailed and precise.
>
> Joshua Meyorwitz, No Sense of Place: The Impact of Electronic Media on Social Behavior, Oxford University Press, 1985 p. 35.

Tengis Baasanjav was born in Ulan Bator, Mongolia, with a heart condition that, if untreated, is always fatal. Doctors in Mongolia did not have the equipment and expertise to address Tengis's condition, but the baby's uncle, who was working in California, sent emails to cardiologists throughout the United States seeking help for his nephew. Dr. Evan Garfein, a Harvard University medical resident at Children's Hospital Boston, spotted the email, and the words "pediatric cardiologist" caught his attention. He brought the case to the Chief of Cardiology, Dr. Pedro del Nido, who agreed to waive his fee and operate on Tengis. There were hurdles before surgery could take place. Dr. del Nido needed to see an echocardiogram of the baby's heart to be sure there would not be insurmountable complications. The first set of films, sent by email, was inadequate for a full diagnosis. New films were forwarded to a doctor at the University of Mongolia, who was able to upload the echocardiogram to Tengis's uncle in California, who created a CD that he sent overnight to Children's Hospital. With the cooperation of several agencies helping to fund the Baasanjav family's trip to the United States and cover the hospital costs, Tengis was successfully operated on. A year later, he was walking around the family home like a normal child. (1)

You've Got Mail!

Digital technology can miraculously unlock communication channels and re-shape the conduct and delivery of healthcare to save lives and deliver better, safer medicine. Without email, Tengis's story would never have reached the U.S. doctors who cured his heart problem. On a daily basis, you are online communicating with business associates, family, and friends. You take for granted that you can find an ATM anywhere in the world to retrieve cash from your bank account. Right now, healthcare does not work that way. You cannot access your health record from anywhere in the world. This limitation still exists, despite the goals set by the Institute of Medicine in 2001 when they mandated that:

The healthcare system should be responsible to patients at all times, twenty-four hours a day, every day, and access to care should be provided over the internet, by telephone, and by any other means. (2)

Conventional email is the most common form of digital communication. It is inexpensive or free. It is something you use every day. However, emailing your doctor may not be met with a positive response. Several patient surveys indicate that people would like to communicate with doctors via email, but only a tiny percent (under 10%) are able to do this because doctors do not use email with patients. Many respondents to these surveys also reported that the use of email influences their choice of a doctor. (3)

Janice lives three hours away from her doctor. She was mowing her lawn on a Saturday when she was bitten on her hand by a white-faced wasp. The hand swelled up and hurt. Janice called her PCP, who is on the staff at the Beth Israel Deaconess Medical Center in Boston (BIDMC). He told her to take a digital photo of the bite and send it to him via email. The doctor wrote back to her within a few minutes after receiving the photo and told her to ice the hand, elevate it, and watch it for twenty-four hours. If it was better, he said, do nothing; if it was worse, she should call him. She would probably have to go into the hospital. In twenty-four hours, Janice sent another email with a new photo, and sure enough, the hand looked better, the pain had subsided, and she was on the mend.

Email can save you from unnecessary visits to the doctor's office or the hospital in nonemergency situations. Because email extends your care beyond the office visit, you can use it to ask those pressing questions that you forgot to ask, did not think about, or were embarrassed to ask when you were face-to-face with your physician. The result is a more comfortable relationship. Email enables you to attach photos, and both the photos and the text of your email become part of your written record, providing a reference point and a history of your interactions with your providers.

Richard is an engineer with high blood pressure who checks his pressure at home daily and emails the readings to his doctor on a spreadsheet at the end of each week. An assistant in the office retrieves the spreadsheet and graphs the numbers, which clearly show that Richard has hypertension with a lot of variability. With this information, Richard's doctor is able to adjust his medication and get the blood pressure completely under control without the time and expense of an office or emergency room visit.

Email use is particularly effective in treating patients with chronic conditions who need to stay in close touch with their physicians. There are many successful examples of patients with ongoing health issues, including diabetes, hypertension, asthma, and congestive heart failure, who email daily or weekly readings of their blood pressure, blood sugar, and peak flow meter to their physician's office. With this updated information, your doctor can regulate your medications, track your progress, and supply you with direct links to educational information on the web that will help you understand and manage your illness.

Email also spans geographic distance and enables you or your doctor to engage specialists on your behalf who might not be local. With the transmission of text and graphic images, email shortens the miles between you and your PCP when you are traveling.

Deborah was on vacation in the Southwest when she injured herself climbing. She saw a local physician, who diagnosed a tendon dislocation and recommended im-

mediate surgery. Unwilling to undergo surgery in an unfamiliar hospital, she sent an email with a photo of the injury from her smartphone to her PCP, asking whether or not she could wait for surgery until she got home. Her PCP set up a consult with a local orthopedist, who reviewed what had happened and told her it was okay to wait. When Deborah arrived home two days later, everything was in place for her surgery, which was successful.

Email Benefits

There are many good reasons why email communication is beneficial and fosters a healthcare environment where continuity of care is a priority. With email, your experience with your providers begins long before you walk into the office and continues long after you return home. Among the benefits of email:

- Email takes real-time constraints out of noncritical doctor-patient communications and enables you to have an ongoing discussion, where you and your doctor exchange information asynchronously to clarify all of your issues and explain all of your concerns.
- Whether it is regulating diabetes, maintaining weight loss, overseeing your blood pressure, or handling stress, email feedback between you and your providers offers the incentives needed to keep you on track.
- Email allows you to send graphical images so that wounds, rashes and other external injuries can be viewed, reviewed, and resolved.
- Email provides the comfort of knowing that no matter where in the world you might be located, if you become ill, you can connect with your own physician, who will oversee your medical care and ensure that all of your conditions and issues are being considered.
- Research on weight reduction has found that people who get dieting feedback via email lose more weight than people who diet without that assistance. Similar studies have shown that people who get email reminders better adhere to their medication schedules than those who are not reminded, and people who receive emails reminding them that they have a scheduled appointment are more likely to show up. That is because email provides the continuous communication and reinforcement needed for behavior modification and lapses of memory.

Privacy and Etiquette

Your medical providers may refuse to engage in email communications out of concern for the privacy of your health information or the fear that you will overwhelm them with insignificant, meaningless emails that take time out of their busy day, time for which they are not reimbursed under the current system. Privacy of your health information cannot be guaranteed when email is sent without proper encryption. Most email systems have basic encryption applications built in that you should use whenever you are communicating with your providers. There are many documented cases of individuals who have sent email to their physicians through company servers. Those emails have been intercepted, and the senders have experienced repercussions related to job advancement. In many companies, sending or receiving personal email via the company server is not allowed; therefore, the company has a right to intercept and view the emails.

Physicians also worry about HIPAA Health Insurance Portability and Accountability Act) regulations that outline their obligations to insure the privacy of your health information. HIPAA, which is discussed in detail in chapter 9, is legislation that defines and restricts the communication of your personal health information by your providers in order to protect the privacy of your health information. Email interactions are not directly addressed by HIPAA. As an e-Patient, you must be aware of privacy risks yet still advocate for email use for nonemergency issues. It is also up to you to ensure that you have access to the full portfolio of communication pathways to your physicians while at the same time overseeing that your communications are secure. The guidelines below address many of the negatives associated with email.

- When possible, email should be sent through a secure web portal site (described on the next page) so that your privacy is protected, unless you are dealing with an issue where privacy is not a concern or a portal is not available.
- You should always encrypt your email using the security settings that your email system provides.
- All emails you send should include your full name, telephone number, and medical or hospital number so that you can be properly identified.

- Your emails should be concise and to the point, never trivial.
- You should always review an email before sending it to avoid any ambiguities.
- You should use email only for standard clinical questions, nonemergencies, and in instances where you might be able to send a digital photo. Do not use it for time-sensitive requests or issues that you wish to be kept confidential.
- Email communications are added to your medical record.
- You should have agreed-upon expectations with your physician that email communications will be answered within a reasonable span of time—twenty-four to forty-eight hours, not two hours or two weeks.

Patient Portals

Patient portals provide the secure communication platform needed to enable private email communications. Many portals also provide you with a view into your electronic health record, and they support ancillary functions between you and your healthcare provider (e.g., prescription refill requests, appointment requests, electronic case management, and lab results). Your patient portal is typically located on an internet site designated by your provider organization, health plan, or employer, and enables you to interact and communicate with your healthcare providers. Portal services are typically available on the internet twenty-four-seven. At a minimum, patients using portals are able to review their bills, access insurance and admission forms, refill their prescriptions, schedule appointments, and look at educational tools such as videos on upcoming surgery or procedures, tip sheets on nutrition, and CDC warnings on public health threats. More robust portals offer e-messaging (email) and e-visits (asynchronous conversations with your providers). You generally have to sign in to your portal each time you return. All portals are required to be HIPAA compliant.

Jack Caine, a forty-four-year-old man traveling on business, suffered a seizure and was hospitalized for several days in an institution far from his home. Using a Relay Health portal site, his doctor was able to provide key medical information to the treating physicians and keep Jack's wife informed throughout his hospitalization. The communication between the distance doctors and the primary care physician led to a speedy recovery and much reduced stress for the family.

With a portal, your providers can send you messages such as appointment reminders, electronic statements, and notifications asking you to check your lab results. Portals have the potential to increase your involvement in your care and can be especially helpful to people with chronic conditions who use the portal to post their blood sugar, blood pressure, and other pertinent personal health information each week. An individual in the physician's office assigned to check patient information on the portal helps to ensure that an irregularity will be addressed before it escalates into the need for a visit with the physician or, worse, to the emergency room. Healthcare organizations are also using portals to send alerts to patients about clinical gaps in care and about scheduling exams, flu shots, and other necessary reminders.

The portal has brought a level of efficiency to patient support unattainable by the older telephone contact systems. At Memorial Sloan Kettering, where a patient portal (MYMSKCC) was launched in 2008, the most popular features among its seven thousand users are scheduling, test preparation information, and e-messaging. Patients preparing for a CT scan can find out exactly what procedures they need to follow before they come in. For those patients who use e-messaging, emails are routed to the appropriate administrative or nursing staff based on the message topic selected by the patient from a predefined list. With the click of a button, the message becomes part of the patient's electronic health record. (3)

On the Duke University portal, Health View, healthcare providers have shared millions of test results with patients since launching in 2008. The Health View portal is designed to serve more than a half million patients using the three major hospitals in the system. Patients, who must register to use the site, are able to access personal health information, billing and insurance information, and forms needed prior to medical visits, as well as pay their bills and schedule nonurgent appointments. (4)

George is a seventy-two-year-old Air Force veteran who used to rush to the VA hospital in West Haven, Connecticut, as many as ten times a month for crises caused by his diabetes, multiple sclerosis, and advanced heart disease. Today, George manages much of his care at the computer, where he posts his blood sugar, heart rhythms, and blood pressure daily to the MyHealtheVet patient portal. A Veterans Affairs nurse at the nearest VA hospital checks each day to see if anything is concerning. George was hospitalized four times. If there's a problem, he can't handle by adjusting his diet or exercise, he has an e-visit with his doctor. (5)

Smartphones

The telephone is your communication device of choice in an emergency. Ironically, health information technology had its roots in the invention of the telephone: chief among the early subscribers of telephone service were doctors and druggists. Prior to the invention of the telephone, you or someone in your family had to physically go to the doctor's office to alert the doctor of a problem. Once the telephone was commercially available, it became a reliable mode of communication, and the doctor did not necessarily have to make a house call or have you come in for a visit. The telephone is still used extensively for patient communication. Sometimes however, telephone delays and telephone tag make it difficult to get the answers you need in an efficient way, given the technology now available. Some of the newer options for communicating with your clinicians include:

Text messaging which always must be under secure conditions has become a more common method for communication between you and your doctors. There are many new text messaging platforms and apps that enable secure text messaging, either through a patient portal or on directly with specific software packages. e-Patients should avail themselves of this.

Smart message systems, (SMS), were devised as a way to send messages to a smartphone and in healthcare are used to notify you of when and how to take medications, by physician's offices to send an appointment reminder, a general alert, or any type of important message. SMS generally works via a one way not a two-way communication path so you receive but cannot send SMS notifications.

Sarah is a sixty-two-year-old woman with multiple medical issues requiring her to take a daily dose of medicine that suppresses her immune system. Sara developed a painful rash on her body. She called her physician's office and reached an answering machine. The next day, Sarah called again and connected with the office receptionist, who told her that her doctor was on vacation, but the on-call doctor would be back to her by the end of the day. Although Sara stayed at home near her phone, she missed the call back. Minutes later, she could not reach anyone when she called the number left on her voice mail. On the third day, Sara called again and was scheduled for an office visit the next day—four days after her initial call. Sara was diagnosed with shingles, which can be easily controlled if caught early. Because telephone inefficiency, answering machines, and covering physicians who did not know Sara delayed her diagnosis, the shingles spread. Sara required an intensive regiment of medication and experienced a long, painful recovery.

What is wrong in this scenario is the use of an answering machine that provided no direct contact with a healthcare professional and an unresponsive on-call physician who was unfamiliar with Sara's compound medical issues. If Sara had reached a medical professional at the beginning, if she had the option to send a photo of the rash via email or a smartphone, she might have had a faster diagnosis and better outcome. Instead, her condition escalated, and a long recuperation resulted in repercussions and costs for her and the healthcare system.

Frank suffers from sleep apnea. A Bluetooth-enabled smartphone checks Frank while he sleeps. A wireless monitor clipped to his finger or toe overnight collects data on the apnea episodes that occur during the night and sends that information to Frank's doctor over the phone. Frank and his physician analyze those episodes and decide on a treatment.

Today more than twelve billion internet connected devices are in us. Almost every person from New York City to Hong Kong is carrying a wireless telephone or smartphone or tablet. These devices are no longer just convenient gadgets they are packed with a litany of features and have the capabilities of a computer. Smartphones and tablets run over wireless networks through radio waves or satellite transmissions and offer voice and texting. Full-feature smartphones have additional functions, such as cameras, MP3 players, downloadable games, email access, and connections to the internet. As a result, the mobile device that sits in your pocket or briefcase constantly enables you to be online, connected, accessing and sharing information on the go.

Additionally, when you call your physician's office, you are not forced to remain in one place while waiting for a return phone call. Typically, both you and your physician carry a mobile phone. If you have a particular issue that requires your medical provider to visualize your problem, you can send your doctor copies of medical photos, X-rays, MRI and CT scans, images of tumors, photos of chronic leg ulcerations, and various skin conditions directly from your phone over the wireless network. Many phones have built-in trigger systems that set off an alarm to remind you when to take your medications, in what dosage, and when you have missed a dose. You can store your health history in your phone so you do not have to struggle to remember the dates of when you might have had a surgery or a disease. Your smartphone can serve as a communication link from home monitoring devices that send data to your physician, nurse practitioner, or physician's assistant, enabling them to monitor your weight, blood pressure, blood glucose, insulin delivery, heart rate, and body temperature to better manage your chronic issues.

The newer smartphones, run on specific operating systems that enable downloading of thousands of apps similar to those available on your desktop, laptop, or tablet, including: medical, and fitness applications for which you need no special training. There are also apps intended to help you control weight, measure food portions, and adapt exercise routines. There is even an application iStethescope, developed for physicians and used by many patients that monitors your heart rate and enables you to send that information from your phone to your doctor. iStethescope uses the smartphone's microphone to enhance the sound of your heart so it can be heard by your medical provider. This application will become more sophisticated and accurate, until every individual with a cardiac issue could be continuously monitored by a provider's office for better, safer care and outcomes Note, that these applications may not be secure, so you must be cautious regarding what information you store on your smartphone.

With their smartphones, doctors are now plugged into evidence-based medicine, giving them the ability to apply the best available evidence gained from scientific study and discovery to their clinical decisions about how to treat you. This is a giant leap in providing you with the best possible care. At many major medical centers, doctors now carry smartphones to the bedside, where they look up the latest treatments, chart your history, and transmit that information to the hospital's central database. They can also retrieve your records from a central database, beam your information to another doctor for an easy efficient transition when a covering doctor leaves and another doctor takes over, or send a prescription directly to the pharmacy.

One unique feature of your smartphone is the ability to store an emergency contact. Let's assume that you arrive in the emergency room unconscious. If you have entered an emergency contact in your smartphone under the acronym ICE (In Case of Emergency), first responders such as paramedics, firefighters, and police officers would be able to find your contact information in the smartphone and communicate with that individual to obtain important medical information. The ICE program was conceived in 2005 after train and bus bombings in Great Britain landed hundreds of unidentified injured individuals in hospital emergency rooms. British paramedic Bob Brotchie started ICE as a grassroots movement that caught on. It has spread throughout the world's medical community. Participating in ICE could save your life. To program ICE into your cell phone:

- Access the address book feature of your cell phone.
- Enter the word "ICE."
- Enter the phone number of your husband, wife, parent, or friend— whoever would have medical information about you that would be helpful to the doctors in case of an emergency.

Anita was brought into the Montefiore Medical Center, New York City's busiest emergency room, unconscious and alone. Doctors were resuscitating her with no information about who she was or who to call until they found a cell phone in her pocket with an ICE notation in the address book. The doctors recognized the ICE acronym and located her emergency contact, who told them that Anita suffered from Type I diabetes. Having that information, the doctors were able to understand her condition and treat her appropriately. (6)

Smartphones have special hardware features as well, that make them well suited for your mobile twenty-first-century lifestyle. They are small, lightweight, and easy to use and handle, and they fit into your pocket. They automatically turn off when not in use. They are equipped with the ability to manage your personal information, such as a calendar, phonebook, notepad, and email exchange. There are drawbacks to using a smartphone, such as the small screen, the short battery, and rampant privacy issues. Just as you do not want to send an email that involves your health through the public internet, you do not want to transmit private health information randomly through the wireless world.

Email, portals, and smartphones are only as effective as their ability to empower you and your medical providers to communicate, collaborate, and expand participatory medicine to the larger community. It seems that the majority of Americans, rich and poor, well-educated and not so well educated, young and old, have adapted to digital technology for business and social reasons. Over the next several years, rapid innovation in wireless communication will provide even more features and applications to change the way that healthcare is delivered to you.

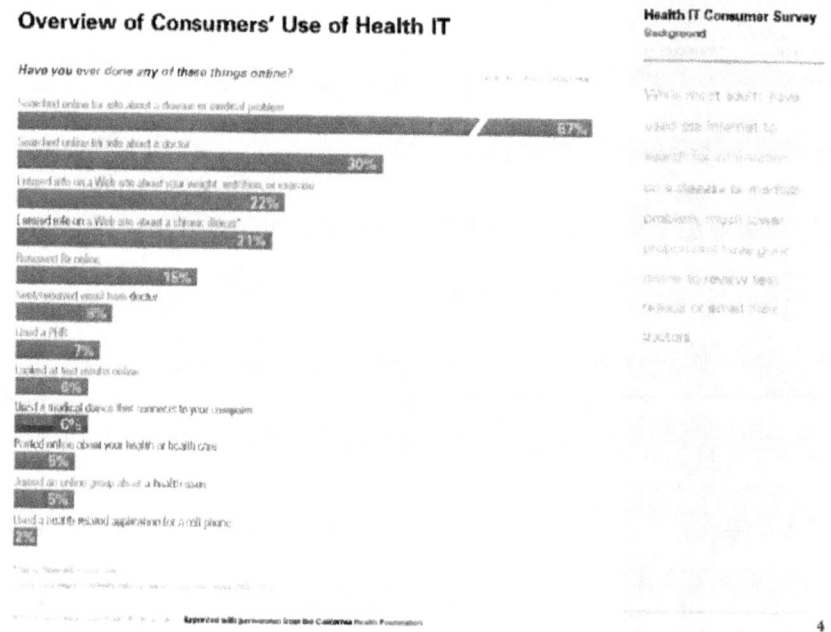

(Reprinted with permission from the California Health Foundation) (7)

Key Points

1. More is better than less when it comes to communication between patients and doctors, as long as it is done with respect and common sense. The currently available tools of communication—email, smartphones, and web resources including portals—are an opportunity for you and your providers to foster the continuous care that you deserve.

2. Many doctors will not voluntarily engage in email with patients, so you have to take the lead and ask for it. There must be ground rules regarding how quickly you expect your emails to be answered and what content is appropriate. You need to understand that email should only be used for non-urgent medical matters and that the privilege of communicating via email with doctors cannot be abused with too many questions or comments that waste the doctor's time.

3. Privacy of your healthcare information is an on-gong problem, and you need to be very aware of who might read your emails, hear your conversations, or see your text messages. It is up to you protect yourself and to decide what information must be kept strictly confidential and when it does not matter so much.

4. The patient portal is the ideal medium where you and your physicians can exchange messages securely and asynchronously. Unfortunately for most individuals, the implementation of a portal is a Provider choice. Once it is available, however, you should make use of the portal for ongoing communications with your provider for scheduling your appointments, retrieving your lab results, and following web links posted on many portals that provide you with helpful health information resources.

5. Those who suffer from chronic illnesses such as congestive heart failure, asthma, and diabetes, can greatly benefit by using email, portals, or smartphones to send daily readings to a designated individual in your provider's office so that you are closely monitored and avoid extra trips to your doctor's office or, worse, the emergency room.

6. As smartphone technology advances, messages, audio reports, and images of wounds, sores, bites, swollen limbs, etc., can be transmitted to your physicians for faster, more efficient solutions.

7. Your smartphone could even save your life if you include your emer-

gency contact (ICE) information in your phone in case you are rushed to the emergency room unconscious and cannot identify yourself to the doctors there.

Notes

1. Ron Winslow and Mei Fong, "Journey of the Heart: A Boston Doctor Struggled to Save a Mongolian Boy by Learning Philanthropic Medicine." Wall Street Journal, April 8–9, 2006. Excerpt with permission from the Wall Street Journal and Copyright Clearing House.
2. Institute of Medicine, "Formulating New Rules to Redesign and Improve Care," Crossing the Quality Chasm: A New Health System for the Twenty-First Century. National Academy of Sciences, 2001, p. 61.
3. Cheryl A. Moyer, MPH; David T. Stern, MD, PhD; Karen S. Tobias, MPH; Douglas T. Cox, MBA; Steven J. Katz, MD, MPH, "Bridging the Electronic Divide: Patient and Provider Perspectives of Email Communication in Primary Care," American Journal of Managed Care 8, No. 5, p. 427.
4. "Online Portal Connects Patients with Treatment Teams," Center News, Memorial Sloan Kettering Cancer Center, May 02009, http://www.mskcc.org/mskcc/html/92607.cfm, accessed July 10 2009.
5. Douglas Goldstein story adapted from a Healthy-Vet case scenario reprinted with permission from the Department of Veteran Affairs and the Markle Foundation.
6. Bob Brotchie, Founder ICE for Safety in an interview with author. http//for Safety
7. Jane Sarasohn-Kahn, MA, MHSA, "How Smartphones Are Changing Healthcare for Consumers and Providers," California HealthCare Foundation, April 2010, http://www.chcf.org.

Chapter Four

e-Patients Go to the Hospital

The best way to survive your stay in the hospital is quite obviously never to end up there in the first place.

Evan Scott Levin, MD, What Your Doctor Won't (or Can't) Tell You, GP. Putnam & Sons New York, 2004.

Evolution of the Hospital

American hospitals have had an interesting history. The first hospitals were makeshift structures used by the military during the colonial era and the Revolutionary War or as quarantine hospitals during epidemics. By the early nineteenth century, sick, poor individuals were treated at alms houses, many of which ultimately became tax-supported municipal hospitals run by local governments. During that era, wealthy individuals were tended at home. Going to the hospital was considered a last resort appropriate only for those who had no other choice.

The first permanent institutions for the general care of the sick were organized in Northeastern cities in the eighteenth and early nineteenth centuries and included Pennsylvania Hospital in 1751, New York Hospital in 1771, and Massachusetts General Hospital in 1821. During the antebellum era (pre–Civil War), the industry saw the growth of many similar general-care hospitals and some specialty hospitals, including asylums or mental hospitals, "lying-in" hospitals associated with foundling homes (unwanted children), and hospitals for the care and treatment of the blind. The link between poverty and hospitalization remained strong during that period.

After the Civil War, the number of hospitals increased dramatically, due not so much to medical advances but to surging immigration and rapid urbanization. By 1900, there were approximately 1,400 hospitals, mostly organized by religious groups or local governments. These municipal hospitals were typically larger than the others and were supported by public funds and managed by public authorities. Many developed reputations for grim conditions. They relied on patient payments to remain financially sound. Finally, there were some private hospitals that derived most of their financial support from philanthropists, supplemented by public funding. As expanding hospital populations outpaced traditional sources of funding, patients at all of these facilities increasingly paid out-of-pocket for their hospital treatment. This helped remove the stigma of hospitalization as a last resort for paupers.

Hospital growth continued unabated in the early twentieth century, with the number soaring to more than six thousand by the mid-1920s. The hospital's image as a charity institution faded, and many hospitals advertised comfortable private rooms to attract patients. The nineteenth-century institutions had mainly treated individuals with severe chronic conditions. These twentieth-century hospitals typically welcomed the acutely ill as well as surgical and obstetrical cases. The number of women delivering their babies in hospitals also increased, especially after 1914, when a sedative to induce "twilight sleeps," that offered the promise of painless childbirth was introduced. This proved crucial to the emergence of the modern hospital. During that early twentieth-century era, nursing schools and medical schools evolved to provide a steady supply of student nurses and medical students to staff wards. Institutional and professional groups formed, including the American Hospital Association and the American College of Surgeons, whose mission was the modernization of all aspects of hospital life, from record-keeping to laundry practices. By the end of the 1920s, deliveries and abortions, adenoidectomies, appendectomies, tonsillectomies, and the treatment of accident victims accounted for 60% of hospital admissions. By 1935, a third of Americans were born in and died in a hospital, but regional and class differences still persisted.

In an ironic twist, high fees charged for hospital care increasingly excluded the hospitals' traditional patients, the poor. Even before World War II, proposals for compulsory national health insurance engendered powerful discussions but did not become law. Meanwhile, voluntary hospital insurance plans, such as Blue Cross, emerged and flourished. The 1946 Hill-Burton Act provided substantial funding through federal grants to states for hospital construction. The states in turn allocated the available money to their various municipalities, creating a network of local hospitals with the intention of achieving 4.5 beds per 1,000 people. Facilities that received Hill-Burton funding had to adhere to several requirements that would shape the hospital environment for years to come:

- They were not allowed to discriminate based on race, color, national origin, or creed, although separate but equal facilities in the same area were allowed.
- Facilities that received funding were required, for a twenty-year period, to provide a "reasonable volume" of free care each year for those residents in the facility's area who needed care but could not afford to pay.
- The federal money was also only provided in cases where the state and local municipality were able to match the federal grant or loan. (1)

Financial issues continued to be central to hospital development. In 1965, Medicare and Medicaid programs provided hospitals with reimbursement for care of the elderly and the poor, which shaped the payment system we have today. In the 1970s and 1980s, as hospital costs climbed, many voluntary hospitals merged, creating regional and national chains. The number of for-profit hospitals rose from 414 in 1977 to 797 in 1997. The 1990s found America's hospitals in a state of financial crisis, aggravated by insurance pressures. As a result, many hospitals shortened patient stays, performed more surgeries on an ambulatory basis, contracted out medical testing, and shifted recuperating patients to less-expensive venues. These changes not only impacted the economics of running the hospital but resulted in raising the quality of care so that there was fewer hospital acquired infections. Most of these healthcare policies remain in effect today.

How Do You Choose a hospital?

There are times when you have an emergency and are taken to the closest hospital emergency room. Although in this situation you do not choose where you are going, once your emergency is addressed, you have the option of being transferred to the hospital of your choice. The actual transfer usually will take place at your personal expense. There are also instances when you know in advance that you are going to the hospital. You want to think carefully about where you are going, who is going to take care of you, and what criteria are important in making your choice.

Here are some tips to help you.

1. **Are you covered?**

   One of the primary factors in choosing a hospital is whether or not your hospital expenses will be covered by your health plan. Hospitalizations can run into several thousand dollars per day, and you need to go to a facility where most of those charges do not come out of your pocket.

2. **Is your doctor on staff?**

   If your primary care doctor is not on staff, and you and your doctor agree a particular hospital is the best place to address your particular needs, be sure your doctor can refer you to someone at that hospital who will handle your overall care.

3. **What type hospital best serves your needs?**

   There are several types of hospitals, and they serve different populations and healthcare issues:

   - General hospitals and community/municipal hospitals deal with a broad range of medical conditions and can adequately address most common illnesses and surgeries. Your local hospital is often more convenient for you and is the place where you will see your primary care physician.
   - Specialized care hospitals address specific diseases or conditions (e.g., cancer centers, children's hospitals, mental health hospitals, and orthopedic centers). They could be the best option if you have a condition that requires the care of physician specialists who are experienced in the unique issues of your particular illness.
   - Teaching hospitals, affiliated with medical schools, are generally larger than community hospitals and handle more comprehen-

sive care issues. At a teaching hospital, you will be seen by medical students, interns, and residents as well as your own doctors. You're in-hospital care could be handled mainly by a hospitalist (a doctor who practices only in the hospital and is familiar with some of the special situations that arise with inpatients) rather than your regular doctor. There are reasons why you might choose your local hospital versus a teaching hospital. The quality of care at a community hospital, especially for routine procedures and surgeries, will likely be as good as it will be at a teaching hospital and will cost less money. On the other hand, if you are in the hospital for a diagnosis that is complicated and rare, the cutting-edge research and technologies of the teaching hospital may be what is needed to offer you the best possible care available.

- For-profit hospital charges are typically higher than the not-for-profit community hospitals for similar procedures. Studies show that there is a higher incidence of adverse events at for-profit hospitals. If you choose a for-profit hospital, you need to investigate the particular institution closely before making a decision. Information that rates all hospitals is available on the web and listed in this chapter on p. 44, 45. (2)

4. **Staff Questions**

You want a hospital that is adequately staffed so that you will get the care you need to get well. Questions to investigate about the staff include:

- How many registered nurses (RNs) are on staff? What is the nurse-to-patient ratio? One nurse can effectively care for three to six patients except in the ICU, where the ratio should be one nurse to every one or two patients.
- Who is the infection control practitioner? The hospital may not have someone with that specific title, but there should be an individual in charge of coordinating infection control, since as many as 5% of patients in the United States pick up a hospital-acquired illness during their stay.
- Who is the ombudsman? Ombudsmen work as go-betweens for

patients and hospitals to handle complaints. Be sure to investigate who is charged with this responsibility.

- How many social workers are on staff? Social workers assist patients in dealing with everything except actual medical issues. This includes emotional, social, clinical, and financial issues, as well as post-hospitalization care plans.

5. **What do people say about the hospital?**

   You need to talk with family and friends about the hospital you are choosing. Ask them about their experience and success in treating their health problem. You can also go online and search for articles, ratings, and blogs about the hospital you are considering. You want to know how many similar cases are seen there each year and the results. If you are having surgery, try to choose a hospital and a doctor that does your specific surgery frequently. If you are particular about the amenities that the hospital offers, check to see if they have what you are looking for. Many hospitals today offer private rooms, massage therapy, and beauty treatments for a fee. If this is important to you, put that into the equation.

6. **Accreditation**

   Be sure the hospital is accredited by the Joint Commission that rates hospitals. Go to: www.jointcommission.org/accreditation/ hospitals. aspx.

   There are also several websites that you where you can check on hospital standings.

   www.hospitalcompare.hhs.gov
Hospital Compare was created through the efforts of the Centers for Medicare and Medicaid Services (CMS), the Department of Health and Human Services, and other members of the Hospital Quality Alliance: Improving Care through Information (HQA). At this site, you will find information on how well hospitals care for patients with certain medical conditions or surgical procedures and the results of patient surveys about the quality of care they received during a recent hospital stay.

www.consumerreports.org/health/doctors/hospital-ratings.htm

The Consumer Reports Patient Ratings are based on survey responses from millions of patients and include how well doctors communicate, how attentive the hospital staff is, and more. The responses are sorted by state and location and include ratings based on how reliably a hospital performs recommended steps before and after surgery to prevent infections and how often patients acquire bloodstream infections while in an intensive-care unit. Hospitals are also compared on how aggressively or conservatively they treat serious chronic medical conditions.

www.qualitycheck.org

This is a website of the Joint Commission on Accreditation of Healthcare Organizations and provides the most comprehensive listing of healthcare organizations available.

www.healthgrades.com

Health Grades rates both doctors and hospitals. In fact, you are able to search information by state and city and then by specialty. On this site, you can find ratings on patient survival immediately after hospitalization and on survival six months after a hospital stay. You can also search for hospitals by the awards they received. Information for these ratings is collected from the Centers for Medicare and Medicaid and from the Leapfrog Group, which seeks to reduce preventable medical mistakes and improve the quality and affordability of healthcare.

7. **Recovery Services**

   With all hospitalizations, there is going to be a recovery period. It could be a combination of hospital recovery care, rehabilitation care, and at-home care. Before you go into the hospital, you need to plan for your recovery period. Seek assistance from hospital staff in finding the right rehabilitation program that will enhance your recovery. Arrange for help at home. Be sure that the hospital has adequate resources to help you with this issue.

What Happens in the Hospital?

Evelyn, a seventy-five-year-old woman who suffered from rheumatoid arthritis and congestive heart failure, fell and fractured her hip. She called 911 and was rushed to the community hospital, where she underwent surgery. After three days, Evelyn was sent to a rehabilitation facility. During her first couple of days there, she made progress and worked with a physical therapist to begin to walk again. On the third day, however, she became weak and tired. Because she complained of pain, her medication dosage was increased by the nurses. The fourth day, Evelyn became unresponsive and confused and could not get out of bed. Family members questioned the staff and asked to have her seen by a physician. They were told that there were no doctors coming in to check the patients until the end of the week (five days later). The family took action, called their internist at the Massachusetts General Hospital (MGH), and with Evelyn's permission, had her transferred. By the time she arrived at the ER at MGH, she was nearly comatose, unresponsive, and hallucinating. The doctors changed and adjusted her medications, and over a two-week period, her physical condition improved and her mental acuity returned. She went to a different rehab hospital to continue the physical therapy that would help the hip heal and where she was taught to walk again. Evelyn lived on for many years after that incident, but had the family not taken action and changed her venue, the medications that were causing adverse reactions could have killed her.

In another example of medical error, three hospitalized babies died after receiving an overdose of the same drug. In the three different occurrences, nurses mistakenly administered a concentration of the medication heparin a thousand times higher than intended, giving these babies a dose with a concentration of ten thousand units per milliliter instead of the correct dosage of ten units per milliliter. (3)

These examples offer vivid illustrations of how easily medication overdose or confusion about medications and dosages can have horrific results. Since these events, drug manufacturers have repackaged many medications and include large "red alert" symbols on the more concentrated doses. Through education and marketing, nurses and doctors have become more aware of drug-drug interactions and the need to closely monitor patients, particularly the elderly. Nevertheless, medication errors are still among the most common medical errors, harming at least 1.5 million people annually, according to the Institute of Medicine of the National Academies. The extra medical costs of treating drug-related injuries occurring in hospitals alone conservatively amount to $3.5 billion a year. This estimate does not take into account lost wages and productivity or additional health issues.

Most hospitals now have bar coding that checks a medication against a patient record and again directly with the patient's ID bracelet to be sure the individual is being given the right medication in the right dosage at the right time. In 2008, the Department of Health and Human Services (HHS) reported that of nearly one million Medicare beneficiaries discharged from hospitals in that month, about one in seven experienced an adverse event and was harmed. These events include infections, falls, and medication errors. The same study showed that over 44% of the adverse events could have been prevented with appropriate attention. (4)

Computer physician order entry (CPOE), where physicians enter their orders directly into a computer to replace handwritten order sheets and prescription pads, helps to avoid some of these medication errors. The built-in decision support capabilities included in most CPOE systems trigger mechanisms warning of an allergy or drug-drug interaction. At hospitals where CPOE systems are in place, medications come to the patient's bedside faster, which can shorten the length of stay. If there is bar coding, medication errors are even more reduced.

A seventy-one-year-old man with congestive heart failure was admitted to the hospital. He did not have a preadmission diagnosis of diabetes. In the emergency room, he had a routine blood test and his blood sugar was elevated. At 11:30 P.M., the nurse notified the covering intern, who ordered an insulin dosage to be given to him. At 1:10 A.M., his blood sugar was checked once again by a different nurse and was elevated even more. The intern ordered more insulin. At 3:00 A.M., another blood test was taken, and the intern ordered more insulin. At 11:00 the next morning, a different covering intern was notified that this man's blood sugar level continued to rise, and eight units of insulin were given intravenously. At 3:40 P.M., the patient was unresponsive, and another blood check revealed a low blood sugar. Later it was discovered that many of the blood specimens had been drawn incorrectly, resulting in a false reading. Fortunately, the patient revived with no lasting harm.

One of the most common causes of in-hospital error happens when new shifts of doctors and nurses are assigned to cover a patient, and full information, including the patient's history and a full assessment of the patient's problems, is not communicated at the point of care. As the above case illustrates, when the medical staff do not have the right information, it is a set-up for medical error. e-Patients in the hospital, should check each medication you are given before it is administered. Ask your nurse about every pill and what is in the intravenous medication attached to your arm. If you do not recognize the medication or cannot get a satisfactory explanation, refuse it until you talk with your doctor. That is your right.

Another issue to be aware of when hospitalized is the prevalence of hospital infection. Hospital-acquired infections have been rising every year. Basic standards for hand hygiene, sterile handling of equipment, and proper use of antibiotics can go a long way in reducing hospital borne infections. You have to keep your ears and eyes open and think about the hygiene around you. Most hospitals have instituted a policy of hand-washing and have special soap dispensers just outside or right inside patient rooms and on all floors. If you do not see your doctor or nurse washing their hands before they examine you, politely insist that they comply with this rule. Make sure that you wash your hands frequently as well, since you come into contact with the germs latent in the hospital equipment, along the corridors, and on hospital personnel. You are the first line of defense.

Starting in October 2012, Medicare payments to hospitals are tied to how well the hospital protects patients from these infections and perform on other patient safety issues. A reporting system adopted by the HHS will enable you to find out how various hospitals perform in relation to similar institutions when it comes to preventing certain infections.

This ruling also addresses the prevalence of in-hospital patient falls. Accidental falls are among the most common safety incidents affecting hospital inpatients and home-care residents. You need to be diligent about asking for help when you have to get out of bed. You may feel the nurses are really busy and you do not want to disturb them, but you do not want to risk a fall that results in broken bones or worse. When you have been ill and are recuperating, you are weaker than normal. With the medications you take in the hospital, you could easily get dizzy and fall. You need to speak up and request help until you are well enough to move about on your own. If that help is not given willingly, speak with the supervisor on the floor or speak to the social worker assigned to your care.

Technology Improves Patient Care

When the doctors at the VA make rounds, they do so with a laptop computer on a portable cart, where they enter notes right at the patient's bedside. The doctors use smartphones that have a database of the patient's bar-coded information, and they match that to the identification barcode on the patient's wristband. All medications also have their own bar-coded identification. Before medicating a patient, a nurse or staff member laser scans three barcodes: the one in the computer, the one on the medication, and the one on the patient's wrist. The scans are entered into the computer. Within seconds, software verifies that medication is for that individual and the dosage is correct. The program screens for half a dozen potential problems, such as drug interactions. If everything checks out, the software simply records the event, preserving the information in an electronic record. The system flashes an immediate warning if anything is out of order. The system makes it very hard to give the patient the wrong medication. The VA has experienced a significant reduction in medication errors.

There is no question that digital communication technology has revolutionized the way doctors, nurses, and other professional staff are able to communicate with each other and with you to provide better, safer care. For example, the use of portable tele-video stations placed at the patient's bedside enable specialists who have similar systems in their home to check on their patients 24/7 In some hospitals, robotic workstations with screens and cameras are visiting patients and communicating visual information and data back to the doctor in his or her office. Smartphones enable doctors to hook into the hospital's larger computer systems to access a patient's electronic record at the bedside and get immediate feedback from the cardiac monitors or from labs while they are in their office or enroute to see you in the hospital. These same doctors can engage in an instant telephone consultation with their colleagues to discuss your care. Nearly 100% of doctors now use smartphones in patient care and patients, too, are using their smartphones to find a trauma center or a hospital when they are on the move and have a medical need.

A Visit to the Emergency Room (ER)

There have been many studies on emergency room usage among patients throughout the country and the data shows the following:

- Number of visits: More than 130.0 per a million individuals.
- Number of injury-related visits: 35.0 per a million individuals
- Average number of visits per 100 persons: 40.4
- Average number of emergency department visits resulting in hospital admission: 16.2 million
- Average number of emergency department visits resulting in admission to critical care unit: 2.3 per a million individuals.
- Percent of visits with patient seen in fewer than 15 minutes: approximately 43.5%
- Percent of visits resulting in hospital admission: approximately 12.4%
- Percent of visits resulting in transfer to a different (psychiatric or other) hospital: 2.3% (5)

The fast response and critical care flow in the emergency room requires your special attention because the potential for error is high. In the ER, the staff is working under extremely stressful conditions, and the complexity of the issues they confront is daunting. In recent years, the number of ERs across the country has decreased while the number of visits has increased. As a result, ERs are experiencing higher patient volume and overcrowding. When you are in the ER, you will wait a longer time to be treated for a minor issue while the more serious cases are triaged ahead of you. That is because ERs triage their patients to provide critical, often life-saving, treatment, which is their primary purpose.

As an e-Patient in the ER, you need to be completely aware of what treatment you are receiving and who is taking care of you, as the staff with its frenetic pace could be prone to making errors. Always try to go to the ER with an advocate (friend or family member). Keep an emergency (ICE) contact in your cell phone. You need to communicate fully all of the information about your medications, allergies, chronic conditions, and special issues, or be sure that your advocate can do this for you so that the providers attending to you have complete information to treat you without mishap.

Nicholas had been experiencing chest pains for ten minutes when his wife made the call to 911. Within minutes, an ambulance arrived and swiftly transported him to his local municipal hospital, where a physician was waiting in the ER to examine him and provide immediate treatment. The Emergency Medical Services squad kept in touch with the ER team using a two-way phone connection. After hooking Nicholas up to the appropriate monitors, taking blood, and talking with him about his symptoms, the on-call cardiologist arrived. Depending on the diagnosis, Nicholas could be sent to the Chest Pain Center for monitoring, to the cardiac catheterization lab, to surgery, be admitted for further evaluation, or be discharged and referred for follow-up care. His wife was allowed to remain at his side during the entire time. Because of the urgent nature of his condition, he was triaged into immediate care services and experienced no wait to be seen. The decision was made to send Nicholas directly to the cardiac catheterization lab.

Several hospitals are using digital dashboards in the ER to track and monitor patients. When patients are admitted, their demographic information, including medications, tests, procedures, and lab results, populate the dashboard. The system provides a real-time view of what is going on in the ER at any particular time, showing updates on each patient. The results are stored, and the dashboard sends alerts to the staff indicating where patients are located and what is happening with them. Patients are identified by their initials to protect privacy, and there are indicators to shows how long a patient has been waiting.

Case Study: The Dashboard at BIDMC

At the Beth Israel Deaconess Hospital in Boston, several staff members got together to create a special workflow system similar to that used by airport traffic control. They use computer monitors to tell them which patients are coming in, who is languishing in a delay pattern, and where all patients are located. The system includes patient admitting information and a full discharge summary, all medication information, and other relevant medical history that is normally on a paper chart. A wireless phone network enables the attending physicians, interns, and residents to communicate with physicians outside the hospital and keeps everyone on the same page in terms of information about the patients. If a patient comes in unconscious but has a driver's license, the staff can access pharmacy information and patient treatment records from affiliated hospitals. A barcode tag system that scans the patient's armband and the tag on hospital equipment enables staff to identify where their equipment is being used and triage it to the most vital cases. Each doctor carries a phone connected to an internal network, an extension of the hospital system, so the doctors can communicate with one another immediately. Thus, physicians can continue their patient care activities and do not have to waste time waiting by a phone for a call back. On the wireless phone system, there is a special number that rings to a senior physician and allows clinicians at other area hospitals direct access for assistance in transferring their most challenging cases. At BIDMC, all prescriptions are electronically stored, so if the emergency room visitor is a patient of the hospital, their medication records are immediately accessible. The dashboard helps the staff make decisions based on the status of individual patients and patient flow and provides information on trends in the ER that can help identify problems. The system has increased efficiency and decreased patient frustration and wait time. (6)

Discharge

When you are about to be released from the hospital, you must be sure you go through a checklist with your nurse regarding your follow-up and treatment instructions. The checklist should include these questions

- Who is your discharge planner? How do you contact that individual if you need information after you are released?
- What medications should you be taking, including dosages, time to take the medication and how to get refills.
- Be sure to understand changes to your previous medication regimen.
- When are you to see the doctor for a follow up visit?
- Are all IV catheters' tubes and patches removed? You do not want to have to return to have those taken out.
- Do you need to talk with someone from the admitting office about any outstanding paperwork, health insurance, or billing questions?
- Who is lining up resources for you to have physical therapy, occupational therapy, or other types of rehabilitation therapy that you might need?

Key Points

1. Although we would like to think of hospitals as safe havens of care, this is not always the case. It is incumbent upon you to be diligent in your hospital choice and aware of the risks and dangers when you are a patient in the hospital.

2. If you are too sick to advocate for yourself, you must have someone with you who can assume that responsibility. That person must be willing to ask tough questions and be assertive if something seems amiss. If you do not have a family member or personal friend to be your advocate, ask the hospital to supply a social worker to work with you.

3. Be sure to ask your doctors what medications are prescribed for you, at what dose, and how often you are to take the medication. Ask the nurses about your medications every time that they are brought to you, or have your advocate do that if you are too ill. This will help to ensure that you are receiving the right drug, in the right dose, at the right time.

4. Inquire about drug interactions. Most hospitals have computerized alert systems. Ask your nurses and doctors if there are any warnings about the combinations of drugs you are receiving.

5. Note when you are given a drug if you are feeling any different—dizzy, lightheaded, nauseous, etc. Don't assume that it is part of being sick. Convey your physical feelings immediately to your nurse, who can double check the drug and call the doctor if you are having a bad reaction.

6. Avoid falls. Do not get out of bed without assistance until you are totally certain that you are able to do so. Wear footwear with rubber soles to prevent slipping. Be sure to report any spills or objects that are on the floor. Use all bars and handles made available to you for balance.

7. Keep personal items, such as medications, drinks, books, the TV remote, and especially the call button near you at all times so that you do not have to get out of bed.

8. Do not hesitate to use the call button if you have a genuine need.

Make sure that you have adequate light to see. If you wear glasses, keep them within reach.

9. If you are on medications that cause you to use the bathroom frequently, ask the nursing staff to schedule time to help accompany you.

10. Observe whether your nurses, doctors, other hospital workers, and visitors use the hand wash dispenser before they come in to see you. Ask everyone coming into your room to wash their hands if they fail to do so.

11. Be sure you understand a proposed surgery. Know exactly what will occur during the operation and who will attend to your care after the surgery, be sure that your anesthesiologist is aware of any problems, allergies, or bad reactions you may have had in prior surgeries or procedures where anesthesia was administered. Be clear about follow-up care when you leave the hospital.

12. When you are being discharged, make sure you get detailed instructions about your medications, diet, and exercise. Also find out what symptoms might be worrisome and ask for a number to call if anything goes wrong. If you will need home care services, a skilled nursing facility, physical therapy, or occupational therapy, get a list of referrals and set up dates for the visits ahead of time.

13. Make sure you leave with a plan for your follow-up care. Be sure you understand what signs and symptoms you should watch out for and what you should do if you have them.

Notes

1. Hill-Burton Program, Hill-Burton Free and Reduced Cost Healthcare (US Department of Health and Human Services Health Resources and Service Administration, July 2010). http://www.hrsa.gov/ hillburton/ default.htm.

2. Gordon Schiff, "Fatal Distraction Finance vs. Vigilance in Our Nation's Hospitals," Journal of General Internal Medicine, April 15, 2000, pp. 269–270. PMCID: PMC 1495441. http://www.ncbi/nlm/ nih.gov/pmc/132358/.

3. Tom Davis, "Deaths of Three Babies in Indiana Spotlight Medication

Mix-Ups," Boston Globe, Associated Press, September 23, 2006.

4. Office of News and Public Information, National Academy of Sciences and Medicine July 2006. http://www.nationalacademies.org/opinnews/ newsitem.aspx.

5. National Hospital Ambulatory Medical Care Survey: 2018 National Summary Tables, table 1, 4, 14, 24, 25pdf icon footnote. (5)

6. Larry A. Nathanson, MD, Director of Emergency Medicine Informatics, the Beth Israel Deaconess Medical Center, in an interview with the author.

Chapter Five

Patient Safety: Ensure Your Well-Being

Mistakes are a fact of life. It is the response to error that counts.

Nikki Giovanni, 1943, African–American poet, essayist, author, and editor William Morris, Black Feeling, Black Talk, Black Judgment, 1970

Josie, an eighteen-month-old baby girl, was admitted to a prestigious East Coast hospital after suffering burns when she climbed into a hot bath. Her mother kept a vigil at her bedside. While in the hospital, Josie became severely dehydrated. Her mother heard the doctor tell the staff that Josie was not to receive any more medication. Shortly thereafter, a nurse came in with more pain medication for Josie. The mother questioned the nurse, who said the orders had been changed. Fifteen minutes later, Josie's heart stopped. She had suffered cardiac arrest as a result of dehydration caused by the medication. Two days later, Josie was taken off life support, and she died in her mother's arms. Her death was preventable. (1)

Medical errors happen far more often than they should. At the beginning of 2001, the Institute of Medicine (IOM) of the National Academies reported to the American public that tens of thousands of deaths occur each year due to medical errors. "At least 44,000 and as many as 98,000 Americans die from medical errors that occur in hospitals each year." (2) "These medical errors kill more people per year than breast cancer, AIDS, or motor vehicle accidents." (3) "These numbers are more than the number of Americans who died in the Korean and Vietnam wars combined, as reported by official Pentagon sources." It is a given that human beings, including medical professionals, will make mistakes. Studies conducted by the National Institute of Health have shown that as many as 40% of these mistakes are preventable. (4)

Patient Safety: A Hallmark of Excellent Care

Patient safety problems of all types happen during the course of receiving healthcare. Many are preventable; some are inexcusable; all are unwarranted. These errors are made by good, but fallible, people working with imperfect systems. Today your physicians, nurses, nurse practitioners and physician assistants are administering more potent medications to patients; are required to understand complex and technologically advanced equipment to diagnose disease and deliver treatment; and handle far more information which they must understand and process on your behalf. These innovations to patient care create opportunities for error because the robust infrastructure needed to manage each and every patient is simply not in place. There are many specific causes of medical errors, which account for one-half of all the medical costs annually in the United States. They include illegible hand-writing in medical records, inaccessibility of records, failure to automatically check for allergy and drug interactions, and missing information about a patient at the point of care. Additionally, many Americans are unable to monitor and manage care for chronic illnesses such as diabetes and hypertension, which leads to preventable second heart attacks, kidney failures, painful and debilitating fractures, and other serious but sometimes irreversible conditions.

Specific examples of avoidable medical errors include:

- Adverse drug events
- Transfusion errors
- Wrong-site surgery
- Surgical injuries (such as leaving foreign bodies inside the patient)
- Preventable suicides
- Restraint-related injuries or deaths
- Hospital-acquired or other treatment-related infections
- Falls

- Burns
- Pressure ulcers
- Mistaken identity
- Poor care in rehab facilities
- Order entry errors
- Diagnostic errors

On July 25, 2005, President George H. W. Bush signed Public Law (PL) 109–41, the Patient Safety and Quality Improvement Act of 2005. This law was a response to the Institute of Medicine report "To Err Is Human," which outlined the preponderance of medical errors experienced by the patient populations at the beginning of this century. This legislation set standards of quality, performance, and patient safety that healthcare institutions are now required to meet. The bill includes provisions for disclosure of medical errors. It also outlines a process that certifies which patient safety organizations are authorized to maintain data about medical error and oversee the quality and safety performance of all US healthcare institutions. These standards provide a safety net for everyone. Among the key provisions of the law are:

- The establishment of a center for patient safety within the Agency for Healthcare Research and Quality (AHRQ).
- The establishment of a nationwide mandatory reporting system by state
- Performance standards and expectations for healthcare organizations
- A mandate for the FDA to develop standards for the safe packaging of medications and the naming of drugs so the names do not confuse and overlap
- Provisions for the legal protection of patient safety data so that healthcare providers are encouraged to voluntarily report incidents to appropriate public services organizations

What You Can Do to Protect Yourself and Your Family

There are actions that you can take to ensure the safety of your care:

1. **General Health Oversight**
   - **Develop your PHR (detailed in chapter 2).** You must have a

continuous care record that stays with you from birth to death. You can make this record available to all of your healthcare providers anywhere in the world. This record will provide you and your healthcare team with full access to the right information and resources to address diseases and conditions that you may face. It is the only way to ensure that all of the providers you encounter have your complete information.

- **Find a healthcare gatekeeper.** You need to have a primary care provider who is fully aware of your medical conditions and who can coordinate your care with members of your health team, and who is available to answer all of your personal health concerns.
- **Make sure you have full access to care.** This includes the flexibility to get a second opinion when an initial diagnosis or treatment plan does not seem quite right; the ability to have a lab test or go to a hospital when you feel there is a valid need; the capability to access to your digital medical record so procedures and tests are not duplicated and your results are made available to you.
- **Do not assume anything!** Make sure that all health professionals involved in your care have all the important health information about you.
- **Ask questions.** If you have a test, be sure to get the results and have the results explained to you. Don't assume that if you hear nothing that everything is okay.
- **Learn about your conditions and treatment** by asking questions and checking reliable internet sources. (See chapter 8 for web resources.)

2. **Your Visit with the Doctor**

- Research the physicians in your community or those suggested by your health plan and select a doctor that you will be comfortable with. (See chapter 8 for websites that evaluate doctors.)
- Schedule an appointment with your doctor. Don't just assume that you can walk in and see your physician. If you need a referral, be sure to get one well in advance of the appointment.
- Be sure when seeing you doctors that you bring every piece of relevant

information that would be helpful. This includes all previous health records, X-rays and lab reports, a written, comprehensive list of your medications, your full medical history, your health insurance card, and a list of questions.

- Arrive early, at least fifteen minutes before the scheduled appointment. Often there are forms to fill out before you see the doctor.

- Make sure you and the doctor stick to the purpose of the visit and you do not wander off into discussions about issues that are not relevant to the reason you are there. Develop a plan with your physician regarding how to address new issues following the visit.

- Request handouts or web references for more in-depth information about your treatment or conditions.

- Review the information, recommendations, and medications; make sure you understand exactly what the instructions are; take the full course of medication prescribed; follow-up with the doctor; and be sure that you personally get the results of any tests or address any further issues that arise.

- If surgery or a procedure is recommended, be sure you understand why the procedure is needed, what the risks are, what the success rate for the surgery is, who you can consult for a second opinion, what post-surgical issues you will face including pain management, the expected time for recovery, potential risks and side effects, short—or long-term disabilities, and how frequently that surgery is performed at the recommended hospital. Also check into what type of anesthesia will be used.

- Be sure to have a family member or friend available to advocate for you when you are in the hospital and unable to advocate for yourself.

3. **Understanding your Rights**

- You have the right to fire your physicians if you do not like the way they treat you, if you do not trust them, or if you do not believe they are providing you with the best possible care. You might be limited to choosing a physician within your health plan if you do not want to pay exorbitant rates for your care, but you have the right to make a change.

- Take an active interest in your care and treatment. Ask any question of your physician that you feel is important to you. Do not get hung up on the idea that it is a "stupid" question. You have the right to ask and to receive an honest answer.

- You have the right to be told about alternative courses of treatment, even if your health insurance may not cover them or you may not be able to afford them. If you find that your physician is saying things like, "Well, there is another option, but you probably can't afford it," make it clear that you want to hear about all of the options for treatment.

- You have the right to refuse to consent for any procedure or treatment. If you refuse consent, you may be asked to read and sign a form indicating that refusal.

- You have the right to leave a hospital or care facility against medical advice. You will likely have to sign a waiver indicating that you are doing so of your own free will and against the recommendations of your medical providers.

- If you participate in a medical or drug trial, you have the right to ask specific questions regarding the purpose of the trial, the potential risks, the benefits which you may receive, payments to which you may be entitled, and measures taken to protect your privacy. (5)

Years ago, when you saw the doctor, you were closely followed for one or two prescribed medications. Adverse interactions were rare. Today, many people over the age of fifty are taking at least four or five medications; some may take up to fifteen different drugs daily. More than 41/2 billion prescriptions are filled annually in the United States, from approximately 20,000 prescription drugs. (6) Given the size of these numbers, it is not surprising that medication error is one of the most pressing concerns in healthcare today. The Institute of Medicine reports that medication errors injure or kill over 1.5 million people annually. These critical mistakes are preventable and involve both prescription drugs and over the counter products, including vitamins, minerals, or herbal supplements. Errors occur in all steps of the medication process, including:

- Prescribing errors at the point of care when healthcare professionals send handwritten prescriptions to the pharmacy that are illegible, resulting in misinterpretation of the medication or the dosage when the order is transcribed. Prescribing errors also happen your team of healthcare professionals do not communicate with one another.

- Drug-handling errors when full-strength medications are not diluted properly in the pharmacy.
- Drug dispensing errors, where the wrong medication is given to the patient.
- Packaging and labeling errors, when drugs with similar names, abbreviations, or packaging are confused. Look-alike and sound-alike drug names and abbreviations cause products to be mistaken, resulting in the wrong medication or dose given to a patient. Although the FDA has been working with pharmaceutical companies on this problem, it is complicated and not easily resolved. (6)

In a community hospital in Tennessee, Dr. Hills ordered the medication lithium for Jim, an elderly patient who was hospitalized for bipolar disorder. Dr. Hills handwrote the prescription for lithium that went through the hospital system to the pharmacy. The pharmacist had difficulty understanding the doctor's handwriting, read the instructions incorrectly, and sent up a similarly named drug, Librium, which is used for anxiety and not for bipolar disorder. A patient can safely take 300–600 milligrams of lithium more often and multiple times a day; however, the recommended dose of Librium for elderly people is 5 milligrams two to four times a day. Jim was given 300 milligrams of Librium, 60 times the safe dose. A nurse discovered the mistake a day later, when she noted that Jim was lethargic. Fortunately, he was taken off the medication and survived unharmed.

What You Can Do to Avert Medication Error

- Always keep a list of medications you take. Include the dose, how often you take them, the imprint on each tablet or capsule, and the name of the pharmacy. The imprint can help you identify a drug when you get refills. When your medications change, you must change your list. Enter the changes into your Personal Health Record. Be sure to include over-the-counter medications, vitamins, nutritional supplements, or herbal products you take regularly.

- Keep medications in their original containers. Many pills look alike, so be sure to always be able to distinguish among them.
- Turn on the lights and read the label every time you take a dose to make sure you have the right drug and you are following the instructions.
- Store your medication in a cool dry place out of the reach of children and pets. Be sure that you keep tubes of ointments and creams with medications away from your toothpaste. A mistake could be serious.
- Old medications that accumulate in your bathroom and drawers should be brought to a police station or hospital where there are sealed collection bins that safely store them and are correctly disposed of so they will not pollute the earth or the water. Drugs can become toxic after the expiration date. The shelf life of most drugs is no longer than one year from the date they were filled by the pharmacist.
- Do not chew, crush, or break any capsules or tablets unless there are instructions on the materials that come with the prescriptions directing you to do so. Some long-acting medications are absorbed too quickly when chewed, and other medications won't be effective or could make you sick.
- Use only the cup or measuring device that comes with liquid medication so that you do not chance a dosing error.
- Develop a daily routine for taking your medications and always follow that routine.
- When you get refills of your medications, make sure they look just like the pills you had been taking. If they do not, have a talk with your pharmacist to be sure that you were not erroneously given the wrong medication.
- Make sure all of your doctors know about any allergies and adverse reactions you have had to medicine. Include that information in your PHR.
- When you pick up your medication from the pharmacy, ask the attendant to make sure that it is the drug that your doctor prescribed. Also be sure to take advantage of the opportunity to ask the pharmacist questions about the directions for taking the medication if you are at all uncertain.

Compliance

Caroline is a sixty-year-old woman who loves to cook and spend time with her grand-daughter. She suffers from high blood pressure and diabetes. Caroline sees her doctor every three months and fills the prescriptions he gives her at a local pharmacy. However, she readily admits that she does not remember to take her blood pressure medication unless she develops a headache.

Following instructions about taking medication is complex, but essential. You go to the doctor and are diagnosed with a simple virus and a cough, and then walk away with three different medications. The instructions about how to take each, when to take each, how to separate taking the different medications, which ones need to be taken with food, etc. are totally confusing so that you could easily take the wrong dosage at the wrong time. It is clear that when you communicate with your physician or physician's assistant and work together to map out a treatment plan, you will better understand, follow the directions, and have more success in controlling your illness.

Medication adherence is complex, multidimensional problem and has long been one of the most vexing issues in American healthcare. It is also one of the causes of America's skyrocketing health costs. Studies indicate that in spite of grave concern on the part of our leaders. and many efforts to find solutions, nearly 50% of patient populations do not adhere to their medications and that non-adherence is estimated to cost the healthcare system $100 billion annually. Various interventions, from providing coaches to mentor patients to help them establish good habits and manage their health conditions, to providing patients with various tools such as pillboxes, physical and virtual, that include bells, whistles and other intrusive noises that pop up, has not made a significant difference

A national survey of 1,020 adults (age 40 or older) with chronic conditions, reported the top barriers to adherence as: "Forgot" (42%) "Ran Out" (34%). "Trying to Save Money" (22%)". Other barriers: "Didn't like taking It:" "Wasn't working;" "Had side effects;" "Didn't think it was needed," make up the rest. (7) Knowing that we live in an era when efficacious drug therapies exist and new ones are being developed at a rapid rate; an era when effective intuitive and smart apps are being developed and refined to help keep patients and caregivers on track with medication adherence, it is discouraging that we cannot yet seem to move the medication adherence needle beyond the 50% mark.

Do the new digital pills and Medisafe® apps make a difference? Somewhat, but they will not resolve the problem, because this is not a technology issue, it is a people issue. It is an issue for clinicians who must be more collaborative and diligent in explaining and ensuring that their patients understand why they need a medication and how to take it properly. It is an issue for patients and their caretakers who must commit to a treatment and follow through. It is an issue for the healthcare system that must find better ways to help those patients who do not have the health literacy to understand the importance of their treatment protocols, or who cannot afford their medications and thus do not fill their prescriptions. Until we can solve the "people" issues, all the apps, bells and whistles in the world are not going to resolve this problem.

The New England Journal of Medicine reported that in 2005, 33–69% of medication-related hospital admissions in the United States were due to poor medication adherence with a resultant cost to the American public of over $100 billion a year. The article indicated that among patients with chronic conditions, adherence to a medication regimen drops radically after six months. Among patients taking antidepressants, over half of them stop taking the drugs three months after the initiation of therapy. (8)

You might fail to take prescribed medication correctly because of forgetfulness, confusion about how to take it, unwillingness or inability to assume the cost of the prescription, concerns about potential adverse reactions, previous experience with adverse reaction to medications, lack of information or understanding about how to use a specific remedy, or assuming that because you are feeling better, you no longer need the medication. Some of the ways to avoid non-adherence are as follows:

- Tell your doctor you want to know the names of your medications and the reasons you are taking them. Make sure that you have full understanding before you leave the physician's office.
- Discuss with your physician the cost of your medication and the financial burden you face to see if there are alternative treatments that you could better afford.
- Request information about your medications or go to the internet and research the drug.

Electronic Prescribing (E-prescribing)

Ellen is a fifty-year-old woman who takes six different medications for mild depression, reflux disease, and chronic asthma. At her annual visit, Ellen and her doctor review her medications. The doctor makes adjustments and, with a new e-prescribing system, sends Ellen's prescriptions directly to a pharmacy located a couple of blocks from her home. Thirty minutes later, Ellen picks up her medications, packaged with comprehensive instructions. She asks the pharmacist a couple of questions and leaves feeling completely informed about what she has to take, when, and how often.

Since there are nearly four billion prescriptions issued annually in the United States, prescribing medicine is a physician's most frequently used tool for addressing patient problems. The proper distribution of these drugs has a profound effect on the outcome of your treatment. E-prescribing is defined as entering a prescription into an automated data entry system (handheld, PC, or other) and generating a prescription electronically. It is based on computer technology that has the capability to enter, modify, review, and communicate drug prescriptions between the doctor or nurse practitioner and the pharmacist. In more sophisticated e-prescribing systems, the computer uses its built-in clinical decision support to check which medications you are currently taking, whether or not a new prescription is covered by your health plan, what possible drug-drug interactions might occur, whether or not the dosage is appropriate, whether or not you have a history of allergies to anything in the drug being prescribed, and any other factors that are relevant in the administration of your medication.

When you arrive at the pharmacy to pick up a prescription, detailed instructions on how to take the medication are included. E-prescribing speeds the process of dispensing medication and provides the cross communication among you, your doctor, and your pharmacist. It results in safer medical practices and helps control costs by prompting your doctors to select generic drugs that are less expensive.

E-Prescribing has continued to raise the bar for safety in dispensing prescriptions in 2021 with better data quality and more efficient communication between pharmacists and prescribers. With 20% to 26% of US adults using telemedicine in 2021, it's no surprise that E-Prescribing use grew throughout the year because approximately 100,000 prescribers added E-Prescribing into their routine, increasing the total number of prescribers using E-Prescribing by 10%. They joined virtually every pharmacy in the US in handling prescriptions electronically.

Online Pharmacies

Diane is fifty-seven, and she lives alone in a small home in Burbank, Iowa. Although she is mobile and able to care for herself, she suffers from diabetes and hypertension and requires five medications daily to keep her healthy. Her doctor arranged to have her order and manage her prescriptions through a state-licensed online service that has a digital copy of her medication profile, insurance information, and conditions and allergies. An online pharmacist communicates with Diane by phone and email, and she is able to ask questions and access internet resources that he recommends. All of her pharmaceutical products are mailed to her so she does not have to leave her home.

Online pharmacies provide an invaluable service to patients who are elderly, disabled, or live in remote areas. Those online systems used by insurance companies save you money for three-month orders and are generally as reputable as your local pharmacy. However, some health websites are outright scams. In late 2005, US federal drug investigators shut down 4,600 illegal internet pharmacy sites and arrested eighteen people. Many of these sites purport to offer prescriptions by legitimate doctors following "safe and secure" online diagnoses that typically take the form of a short questionnaire. The scam sites lead the unwary subscriber to believe that their health form is reviewed by a physician before being sent on to a pharmacist, who fills the prescription. In reality, many of the websites fill the orders or send them to an illegal wholesaler without any healthcare professional involved.

There are ways for you to save money on prescription medication via online resources. The Veterans Administration enables veterans to refill prescriptions online as a part of the VA's MyHealtheVet service. Five states—Illinois, Kansas, Wisconsin, Vermont, and Missouri—have implemented an online drug program, I-SaveRx, that allows citizens to purchase cheaper, safe prescription drugs from state-approved pharmacies in the United States, Europe, and Canada. With the exception of these few programs, there are hazards to online medication purchases, and before using them, you should check the FDA guidelines regarding the use of online prescription fulfillment services at www.fda.gov/buyonline. Other Resources for Safe HealthCare and Medication Advice include:

www.ahrq.gov/consumer/5steps.htm
www.fda.gov/medwatch/how.htm2.
www.ismp.org/pages/consumer.html
www.medmarx.com
www.nabp.net

Key Points

1. Resources are now available on the web to check the credentials of physicians and the ratings of hospitals. You need to choose a doctor you are comfortable with, who exceeds your expectations when you are ill by being available and answering all of your questions. This is a matter of your medical safety and not just a whimsical desire. If you have to go into the hospital, you want to choose an institution that has an exemplary record for patient safety and a long history of handling the procedures or surgery that you face.

2. As an e-Patient, you can take responsibility for your health by fully participating and, for example, creating a personal health record that is available to all of your providers and asking questions when you do not understand a treatment or medication.

3. Medication error is the most egregious shortcoming in the medical system, and both patients and providers are responsible. You must always have a list of all your medications available when you are seeking care. You must take it upon yourself to understand the side effects and your allergies to medication. You must be organized about how you store and administer your medication and must adhere to the treatment plan that you have agreed upon with your physicians.

4. You have the right to know what is going on with your health, including results of tests. If you do not understand something, ask your doc-

tor to explain and provide resources for further research, as well as what follow-up is required to resolve your problem.

5. Online pharmacies can serve a valuable purpose, particularly for individuals who are homebound or live in remote areas and are unable to access their medication unless it is sent through the mail. However, as with everything on the web, there are good and bad websites. Patients need to be coached in how to distinguish among them.

Notes

1. Sorrel King, mother of Josie King and founder of the Josie King Foundation, in an interview with the author. For more information on the Josie King Foundation, go to http://www.josieking.org

2. Institute of Medicine, "To Err Is Human: Building a Safer Health System," Errors in Healthcare: A Leading Cause of Death and Injury, National Academy Press, 2000, p. 26.

3. Centers for Disease Control and Prevention, National Center for Health Statistics Deaths: Preliminary Data for 1998, 1999. National Vital Statistics Reports, Washington D.C. Department of Health and Human Services.

4. Rosemary Gibson, and Singh Janardan Prasad, "A Million Deaths a Decade," Wall of Silence: The Untold Story of the Medical Mistakes That Kill and Injure Millions of Americans, Lifeline Press, 2003, p. 41

5. Medication Error https://www.fda.gov/about-fda/fda-basics/fact-sheet-f

6. New England Healthcare Institute, "Report on Patient Medication Adherence," Med Ad News, February 2010. http://www.nehi.net.

7. https://cdn2.hubspot.net/hubfs/167852/docs/MTM%20White%20Paper%20vFINAL.pdf?t=1508251339963

8. Lars Osterberg, MD, and Terrence Blaschke, MD, "Adherence to Medication," New England Journal of Medicine, 353 (2005) :487–497.

CHAPTER SIX

Telehealth Expands Provider/Patient Connections

At its very core, our work today is about improving lives and ensuring peace of mind. It's about getting the right care to the right person at the right time—each and every time.

Katherine Sibelius Secretary, Department of Health and Human Services, National Summit on Healthcare Quality and Value, Washington, DC, October 4, 2010.

On November 21, 2004, eighteen-month-old baby Kate was critically injured in a car accident. She was rushed by paramedics to the closest hospital in Douglas, Arizona, a small rural town along the US-Mexico border. Kate was in shock and minutes away from death, having lost almost two-thirds of her blood from multiple injuries. The nearest trauma center was in Tucson, more than a hundred miles away. When she arrived at the Douglas emergency room, the doctor saw immediately that he would need assistance and called the University Medical Center's Level 1 Trauma Center in Tucson. Over the Arizona Telemedicine Program (ATP) network, a trauma surgeon at UMC was able to see the baby and examine her injuries. He reviewed the Kate's vital signs, X-rays, and lab tests results and virtually led the doctors and nurses in Douglas through emergency medical procedures. Once Kate was resuscitated and stabilized, she was transported to the UMC for further treatment. She fully recovered. (1)

Patients are scattered throughout areas of the country far removed from major medical institutions. The Federal government estimates that nearly 80% of residents in rural America are "medically underserved." with fewer than 10% of the country's doctors practicing medicine in those areas. That ratio is worsening each year as the doctors who practice in rural America age and retire, and fewer young doctors opt to work there. Additionally, we know that nationwide, we need sixteen thousand additional primary care doctors to provide basic care for all citizens. Nursing shortages are equally disturbing. *The Washington Post* predicts that 1.2 million new registered nurses (RNs) will be needed by 2030 to address the current shortage. (2)

As a result of these inadequacies in our healthcare system, individuals who live in rural areas of the US, and who need both general basic health oversight as well as critical care assistance, have, for decades, been turning to virtual care, often referred to as telehealth or telemedicine, as the way to provide access to comprehensive health services.

Baby Adam is born at the Hays Medical Center in Kansas with a suspected heart murmur. There is no pediatric cardiologist on staff there; however, at Hays, patients' radiology films are automatically transferred to specialists in Wichita, Kansas, where they are read instantly, and a diagnosis is sent back in hours not days. Within moments of birth, using an electronic stethoscope, Adam's heartbeat is transmitted to the pediatric cardiologist, who is 270 miles away. With a video hookup, he is able to help the Hays cardiology technician do an echocardiogram, which provides the data to make a diagnosis, advise the local doctors on the best possible care for Adam, and reassure the boy's parents that the problem is being appropriately handled. (3)

By 2020 there were more than 12 billion internet connected devices in use The convergence of the internet, high-bandwidth telecommunications, video technology, sophisticated medical robots, and the spread of electronic health records provide the enabling technologies for virtual care, often labeled "telemedicine" or "telehealth." to work. Virtual health connects patients with doctors, specialists, therapists nurse practitioners, nurses and other healthcare providers. who can be geographically distanced, by hundreds or thousands of miles? Using interactive audio and video conferencing, as well as high-speed telephone lines that transmit and send lab tests, X-rays, and vital signs, patients can and do receive diagnoses and treatment plans. Video technology that enables face-to-face examination allows you to receive physical therapy, occupational and mental health treatment, and hospice services remotely, as well as services for homebound individuals to monitor chronic conditions hourly or daily. Much of this was happening in rural areas of the country before the pandemic. Now telehealth has been expanded throughout the country as a result of the passage of the Coronavirus Aid, Relief, and Economic Security Act, (CARES) that includes provisions to broaden coverage of, and provide grants to support the greater use of telehealth services,

In many remote areas including Arizona, the Dakotas, Idaho, Montana, Oregon, Washington, and Wyoming, telehealth technology enables patients to access needed medical services, as the following examples illustrate.

1. Dozens of asthmatic children are using web cams in their homes to send videos showing how they use their inhalers and send readings from their peak flow meters (airflow measurement devices) to pulmonary specialists located at major medical centers.

2. Diabetic patients on Indian reservations in rural Arizona now have their retinas screened and treated remotely by ophthalmologists at world-renowned eye centers throughout the country for diabetic eye changes that could lead to complications and blindness.

3. Telehealth also has other benefits, such as enabling emergency care and keeping family members in the loop when patients are transported to distant hospitals.

Twin babies born in rural Washington State went into distress shortly after their birth and were transported immediately by helicopter to the closest large neonatal care center 140 miles away. Their mother had not even seen her babies before they were flown out. Since she had three other children at home, it was virtually impossible for her to go with her newborn twins. Northwest Telehealth, a telemedicine network of Inland Northwest Health Services in Spokane, Washington, arranged for a telephone connection with video screens between the mother and her babies. For six weeks, using a telehealth camera, the mom could see her one-pound babies, watch them receive a feeding, hear them burp, and enjoy their first smiles while tending to the immediate needs of the others in her family. This bonding could never be replaced after the fact.

The following are other examples of how telehealth has been used throughout the country to enable life-saving virtual connections over a long-distance.

Teletherapy

Bob is a farmer who lives in rural Minnesota. He was severely burned on his hands, chest, and back while fixing some farm machinery. He was treated initially in a local hospital and, after he was stable, transferred to the Regional Burn Center in Rochester, Minnesota, 450 miles from his home. Once he was released, Bob arranged for his fol-low-up care to be handled over the St. Alexius Telecare network of North Dakota. He participated in three telemedicine visits, each lasting thirty minutes, avoiding a two-day trip that would have necessitated leaving the farm with no one in charge during harvest time This long-distance care enabled Bob to keep the farm going. It also pro-vided relief to the family, because they did not have to transport him 450 miles and back. (4)

Speech Therapy

It was a typical spring day on May 27 for Wes, who worked at a National Guard camp in North Dakota. Suddenly he collapsed while using a mini front loader. He was taken to the local Carrington Health Center, where he was diagnosed with a ruptured brain aneurysm. He was immediately flown to Fargo, North Dakota, where doctors were able to keep Wes alive without surgery. After seventeen days in the hospital and another twenty-seven days in rehab, Wes was able to return home. Doctors recommended that he continue with physical therapy, occupational therapy, and speech therapy; however, speech therapy was not available at Carrington Health Center near his home. With a telemedicine connection from St. Alexius Medical Center, Wes was able to have the therapy, which made him physically more comfortable, and mentally and emotionally more confident. Wes returned to work full time one year after his collapse. Without the therapy, he would not have been able to continue work at all. (5)

St. Alexius is one example of a full-service hospital with seventeen affiliated clinics located in remote areas of North and South Dakota. The St. Alexius telemedicine network has been functioning for ten years, and it enables doctors to see patients who would otherwise not be served by specialists, and in some cases, not even by primary care physicians. Using the network, St. Alexius doctors conduct between four hundred and five hundred consults every year, including 24/7 trauma consultations between the emergency room at St. Alexius and outlying centers. Much of the work they do includes post-operative and post-trauma care, similar to what Bob received. For example, doctors at St. Alexius are able to remotely check a patient's incision while the patient remains in the home. St. Alexius also offers speech therapy to stroke patients, diabetes and Alzheimer support groups, and mental health services that would be unavailable otherwise. Sometimes they even use the equipment to resolve a personal crisis.

There are many other centers of excellence offering virtual care. Marshfield Clinic in Marshfield, Wisconsin, has over 775 physicians in more than 80 medical specialties and subspecialties located in 54 locations throughout northern, central, and western Wisconsin. An integral part of healthcare delivery at Marshfield Clinic is the Outreach Services program. More than 1,200 hospitals, clinics, and other sites participate in these outreach programs. Clinic physicians and staff provide off-site consultation in 52 specialties. The clinic had more than 3.8 million patient encounters for the year ended September 30, 2010, and reported 376,708 unique patients in the clinic system during this same period.

Examples of its services include off-site physician consultation involving more than 80 sub—specialties, twenty-four-hour EKG interpretation via computer, mobile echocardiography, a reference laboratory, regional blood banking, radiology, EEGs, orthotics and prosthetics, radiation physics, pulmonary function, and biomedical electronics.

A telehealth visit at Marshfield Clinic is scheduled by telephone. A specific provider for a patient is selected based on the patient's individual problem. At the appointed time, a trained telehealth nurse-clinician orients the patient, who is at a local clinic, to the telehealth consult process and prepares the patient and family for a clinical exam. The provider and patient can see and hear each other, just as if they were sitting in the same room. The doctor is on TV, and the patient and family are in an exam room with the nurse practitioner, perhaps a hundred miles away. During the exam, specially designed equipment gives the doctor all the information needed to make a diagnosis and to discuss a plan of care with the patient and the family. Handheld cameras can zoom in for tight close-ups to give the doctor a good picture. Digital stethoscopes with headphones that intensify sounds of the heart and lungs enable the doctor to listen to the heart; fiber optic cameras allow doctors to "see" inside the ears, eyes, and mouth. Doctors use a clinical-quality video system on the computer to conduct the consultation. The precision of these instruments allows the doctor to make a diagnosis as if doctor and patient were in the same room. Everyone can talk back and forth during these exams, and questions are encouraged. (6)

Physical Therapy

Physical therapy services delivered remotely are also convenient and efficient for providing services to disabled patients who live in rural locations or are housebound. With available digital technology, there is no need for the patient and healthcare professional to be in the same location. internet-based video-conferencing installed right at the patient's bedside in the home or in a local physician's clinic will enable physical and occupational therapists to assess the physical condition of a patient, communicate remedial movements or remedial sustained postures, and evaluate the patient's progress over time. The therapist is also able to instruct patients to move in a particular manner, to assume a sustained posture, or perform a test. The patient and the distant healthcare provider are able to see each other in full color and hear each other using full-motion video and digital audio systems. This technology has proven successful for acute stroke care and for rehabilitation following surgery or severe injury.

A twenty-five-year-old woman suffered a brain injury and was in a rehabilitation center. Her insurance ran out, and she was discharged to her home in a remote section of Oklahoma. The family was unable to care for her without help and training. They were introduced to a teletherapy program through INTEGRIS Health in Oklahoma City that provides access to medical care for rural Oklahomans and others. A physical therapist worked with family members using phone connections and cameras to teach the patient how to move about. At the beginning of the teletherapy, the patient could not sit in a wheelchair without assistance. At the end of twelve weeks, she was walking around the home with little assistance and able to do a number of life skills for herself. (7)

Mental Health Services

According to the National Institute of Mental Health., about one in five adults, age eighteen and older suffer from a mental disorder in a given year. (52.9 million in 2020). Mental illnesses include many different conditions that vary in degree of severity, ranging from mild, to moderate, to severe. (8)
https://too/health/statistics/mental-illness

There is a severe lack of mental health services throughout the United States, and the problem is especially acute in rural areas. Virtual health appointments by video screen and telephone—are increasingly used by psychiatrists to provide counseling to many who would otherwise be unable to get help. A simple audio telephone connection or video system is all that is needed for a specialist to consult with patients who lives miles away.

A woman from Arkansas, left destitute with four children to support, was clinically depressed and in danger of losing her children. A local nurse contacted a physician connected to the Rural Arkansas Delta Integrated Telehealth system in Little Rock, Arkansas. He arranged for a consultation with a University of Arkansas psychologist. Since she had limited income and no phone or transportation, the local police transported her to a facility near her home where she was able to receive her telehealth consultations with a psychologist over interactive video equipment. With this counseling, the woman was able to function and take care of her children. (9)

Traumas and Emergencies

In trauma medicine, split-second timing can mean the difference between life and death, between the ability to save or lose a limb, between a short-term injury and long-term recurrent problems. With telemedicine providing fast access to trauma specialists, countless people have been saved from prolonged pain, suffering, and even death as they await a diagnosis or a transfer. By enabling trained emergency technicians and nurses to act immediately to control bleeding, administer appropriate medications, suture wounds, and care for patients with assistance from the specialists looking on through the eye of the camera and making the diagnosis or providing instructions, lives are being saved.

Martin, who worked at a lumber mill in northern California, caught his hand and forearm in the chain that moved wood through the mill. A local family nurse practitioner assessed the injury and sent images to the Santa Rosa Memorial Hospital Emergency Department for a telemedicine assessment and consult with the emergency physician and a radiologist. Based on the consult, the nurse practitioner was able to clean, suture, and splint the wound, and Martin avoided a five-hour round trip to the hospital and unnecessary transport costs. (9)

In a unique program forged between fourteen community hospitals and the Massachusetts General Hospital, The Telestroke Service, using trained neurologists, now treat stroke patients via a telehealth network that brings the specialists to the bedside of the patient during those critical first three hours. These specialists work with local physicians to help them diagnose whether the symptoms they are seeing are indeed stroke and the best way to address the problem. Many similar telestroke centers are emerging throughout the nation.

Telehospice: Death with Dignity

Hospice administers a range of health and comfort care services to patients who are nearing the end of life. A much misunderstood and underused service, hospice enables terminal patients to die as they wish at home. Hospice attempts to meet the needs of this population with services such as pain management, spiritual counseling, helping people to get paperwork in order, and providing reassurance to the family. The availability of telemonitoring workstation equipment that is already being used to monitor many homebound also enables patients to transition to telehospice care. There is much technology available today, including telecommunications-ready infusion pumps for pain management and videotapes for caregiver education, that, at the touch of a button, can make a huge difference.

The eICU®: Remote Monitoring for Intensive Care

If you are unlucky enough to land in the intensive care unit, you have cause for concern. There is a severe shortage of ICU specialists, The expertise that an intensive care doctor brings to the critical care unit is missing in many hospitals. Prior to COVID fewer than six thousand intensivists were actively practicing in the United States, leaving only 13% of ICU patients receiving dedicated intensivist care. With an aging population that continues to increase, and its anticipated need for intensive care services, the acute shortage of specialists is a huge problem that was greatly magnified by the pandemic. When an intensive care doctor is available to manage the ICU, studies show that a patient's chance of dying in the ICU decreases by 30%.

The eICU® is a central command room outfitted with cameras that beam directly to each patient's bedside. Broadband technologies and land-based wireless networks allow continuous communication of data, voice, and video at ultra-high speeds. The eICU® can be located within a hospital complex or offsite at a completely different location minutes or miles from the intensive care unit. It is manned twenty-four-seven by intensive care doctors and nurses. All of the information that the doctor needs to make a decision comes directly into the eICU®. From a control station, several computer screens display your diagnosis and progress, doctors' notes, and vital signs such as heart rate and blood pressure. The equipment enables the eICU® doctor to see your medical record, including nurse's notes and blood test results. Video conferencing equipment enables two-way conversation between the eICU doctor and you and your family. The eICU® doctor is also able to zoom in on you and gauge your level of consciousness. This helps the doctor by providing much more information than an average doctor or specialist on call would have. Dedicated hot phones create an instantaneous link between each ICU in the network and the care team in the eICU®. The constant surveillance arms the ICU physician with the patient information needed to make the right decisions. Many large and small hospitals throughout the nation now use the eICU® as a safer, more efficient way to monitor their patients.

Online Second Opinion Consults

A fifty-five-year-old man living in a small town in New Mexico was diagnosed with a growth lodged in his abdomen. There was no indication of whether or not this growth was benign or a cancer. Primary care doctors and specialists all had different opinions as to what to do. One suggested a biopsy, another suggested surgery, and a third wanted to wait. With all of the confusion, the man decided to seek a second opinion and opted to con-tact the Mayo Clinic for advice. Flying to the Mayo Clinic, located in Rochester, Wisconsin, meant time away from work and an expensive trip, so the patient opted for a second opin-ion online via telemedicine. The existence of an electronic medical record, video and audio high-speed transmission, and the ability to send films and lab reports ahead to the specialists at the Mayo Clinic combined to make this consultation possible and successful.

Although telehealth has long focused on long distance connections to pro-vide healthcare in rural areas, during the pandemic telehealth attained much visibility and became a standard of care that fostered patient clinician visits over secure platforms as well as a wider variety of non-clinical and public health services uses

With certain diagnoses, such as heart disease, cancer, autoimmune dis-orders, and other serious illnesses, telehealth enables patients to seek second opinions virtually from leading specialists without having to wait agonizing months and travel great distances for an appointment. Many employers now offer online second opinion consultations regarding their employee benefit packages so when illness strikes you are able to access the best specialists throughout the country to resolve mixed messages and differing diagnoses and help you through the confusion of what to do.

Home Telemonitoring

One out of every eight persons is over the age of sixty-five in the United States. By the year 2030, that number will double, accounting for 25% of the population. These baby boomers face the double challenge of caring for elderly parents while dealing with their own emerging health problems as they approach retirement. With the shortage of doctors and nurses in America, millions of individuals will be able to help their elderly parents, themselves, and their children with the assistance of telemonitoring. Telemonitoring services enable you to check your blood pressure, heart rate, weight, and blood glucose, and then send objective data to clinicians over the telephone or through an internet connection from your home. The wireless systems used in home monitoring are predicated upon a set of noninvasive, wearable monitors (many available for reimbursement), such as the following:

- Blood pressure cuffs connected to a telecommunications plug-in; a similar device to monitor blood glucose that attaches to a telehealth computer workstation. The telehealth work station is connected to your phone line and is linked through the line to a computer at a hospital or clinic, where a nurse monitors the readings.
- A desktop glass ball called an orb that can be set to glow different colors based on preprogrammed instructions when used in conjunction with a smart pillbox. This helps you adhere to a schedule of taking medications by sending a signal from the pillbox to the orb, which glows red when a medication is overdue and green when you are on schedule with your medications. The smart pillbox sends an electronic message to a central server that houses your data. The server then signals back to the orb.
- An implantable cardioverter defibrillator (ICD) that notifies your doctor each time it gives you a jolt to stop a potentially dangerous abnormal heart rhythm.
- A recorder the size of a computer mouse that gathers data from your

pacemaker or ICD. When you hold it to your chest, the device sends the number to a secure computer that you or your doctor can access via the internet.

- A watch that checks your blood sugar levels through the skin and transmits the readings to a remote site, where a nurse or nurse's assistant is monitoring the input.
- A soft machine-washable undershirt that can take a continuous movie of your heart's electrical activity, blood pressure, oxygen level, and other health indicators.
- A ring the size of a college class ring that continuously measures body temperature, heart rate, and oxygen levels in the blood and sends this information to a cell phone or computer.
- A wireless electronic bathroom scale that sends weight data into a home monitoring system (a device the size of a phone answering machine) programmed to flash if the patient forgets to check in.
- A similarly constructed wireless blood pressure cuff that sends blood pressure readings to the device, which then sends these readings directly on to a computer at a doctor's office or clinic, where a nurse or case manager will review it.

To keep in touch with his nurses, eighty-year-old Ed has been using a home monitoring system with two-way video for many years. Each morning, he steps on a mat that records his weight. A bracelet sensor reads his blood pressure, heart rate, and oxygen levels. These indicators are sent via a telephone connection to a group of nurses at a remote location, and if there is an irregularity, they video teleconference with Ed. Using an electronic stethoscope, they can listen to his heart and talk with him about what might be wrong. If the problem looks serious, Ed is advised to go to the hospital. The home monitoring crew alerts the hospital staff that he is coming and provides them with the details of his condition. Most times, hospital visits are not necessary, and all that is needed is a simple adjustment of Ed's medications.

With remote phone or video consultations such as those that Ed uses, every patient who is located in a rural area where medical services are not easily available, as well as patients who live in urban areas but are homebound, can self-manage many of their health issues. The technology also enables your providers to remotely identify the onset of complications and give you an action remedy right away or determine that you must travel the distance to come in to see them. When you do get to a physician's office or the emergency department, your consistent home monitoring provides the baseline information for a quick assessment of what is wrong. All of this eliminates many of those costly, frightening emergency room visits.

Hays Telemedicine in Hays, Kansas, conducted a study on congestive heart failure in patients who were monitored at home by telemedicine nurses. These patients were seen weekly using a modified television in the home connected to a television cable system base station. At the base station, a telemedicine nurse monitored the study participants. Home and office visits were prescheduled, and when a home intervention was required to draw blood or do other lab tests, someone from the Visiting Nurses Association made the home visit and collected the necessary data. The study showed that the patients had 20% fewer emergency room visits; they were more likely to remain in their homes instead of going to a nursing home; and there was a significant cost savings to the individuals and to the system. (10)

Home Monitoring Systems

There are several criteria for you to be considered for a home monitoring system that is reimbursed by health insurance, Medicare, or Medicaid:

- You must have a disability.
- You must have impaired ability to care for yourself.
- You must be recovering from an illness, surgery, or hospital stay.
- You must be living with a chronic illness that is poorly controlled.
- You must have a terminal illness.

Additional home services range from skilled medical care to home support services, or a combination of both. Skilled care is directed by your doctor and is provided by healthcare professionals such as nurses, medical social workers, and physical therapists. It can include activities such as home dialysis, drawing blood, administering medication, and physical and occupational therapy. Home support services help you with basic everyday issues such as dressing and bathing, and are provided by home health aides.

Finding Home Care Services

The first step to finding a home care service is to talk to your doctor or other healthcare professionals, such as physical or occupational therapists or social workers. If you are looking for a reputable home healthcare agency, a hospital social worker or nurse may be able to help. Other sources include your state and county health department, local agencies on aging, and the national Eldercare Locator, a public service of the Administration on Aging. Once you find a home healthcare service that you think will satisfy your needs, questions to ask about their qualifications to assist in making a decision include:

- How long has the agency been in business?
- Is the agency approved by your health maintenance organization or supplemental insurance?
- Is the agency evaluated and accredited by a governing body such as the Joint Commission's Home Care Accreditation Program or Medicare?
- Is the agency licensed by the state? (Most states require agencies to be licensed and reviewed regularly, and the report is available upon request through the health department.)
- Can the agency provide references (a list of doctors, hospitals, discharge planners, and former clients)?
- What are the credentials of the providers who work with this agency?
- Does the agency make clear its policies regarding billing, services, and fees?
- Does the agency provide any resources that will help you with financial assistance for their services?
- Are there days of exclusion before your coverage will begin?
- How many hours of coverage each week are allocated, and is there twenty-four-hour coverage for emergencies?
- Does the agency involve you or your family in planning your care and making adjustments? Do they require involvement of a family member and specify what is required of that individual?
- Is there a program for conflict resolution and a clearly defined process for answering your questions or complaints?
- Do they provide and service special equipment such as oxygen, respirators, dialysis, and train you in how to use the equipment?

Among the ways that home healthcare services are reimbursed include:

Medicare: Medicare may pay for medical home healthcare services through a certified home healthcare agency if a physician orders this service. Services covered by Medicare include skilled nursing assistance or physical, speech, or occupational therapy. If your home health services are covered under Medicare, your doctor, care manager, or discharge planner will probably make arrangements for a home healthcare agency.

Medicaid: Depending on your income and assets, anyone is eligible.

Older Americans Act: This federal program funds state and local programs that provide services to frail and disabled individuals who are sixty years old or older.

Veterans Affairs: If you are a veteran and at least 50% disabled due to a service-related injury or illness, you may be eligible for medical services through Veterans Affairs hospital-based home care services.

Community Organizations: Depending on your situation and finances, certain community organizations cover home care costs.

Insurance: Many insurance programs cover some home healthcare services for short-term medical needs. However, long-term coverage varies. Long-term care policies are available through private insurance companies and the federal government.

Telehealth Costs

One of your biggest healthcare concerns today is the cost of your care and how to keep it under control and within your budget. Virtual visits played a key role in solving some of the cost issues built into face-to-face delivery of medical care, that has been the norm. A major benefit of telehealth is the reduction of the number of patient visits to doctor's offices, clinics, and hospitals. Those visits could involve the transfer of individuals from nursing homes, correctional facilities, and local hospitals to trauma centers. Not only are these transfers expensive, but sometimes they are wasted effort. Many of the issues that require transport are easily handled by home monitoring and telephone consultations. Monitoring the vital signs of individuals with chronic disease such as congestive heart failure and diabetes has the potential to reduce the country's healthcare costs by several billion dollars over the next twenty-five years. Those savings would come from spotting problems before they develop, thus reducing emergency department visits, preventing hospitalizations, and cutting the length of hospital stays.

Research from the Center for Information Technology Leadership (CITL) confirms that telehealth saves money. Their study, The Value of Provider-to-Provider Telehealth Technologies, examines the cost-benefit of using virtual technologies, such as real-time video and store-and-forward data capture in emergency rooms, correctional institutions, nursing homes, and physician offices. Their report concludes that if a telehealth initiative were nationally implemented for a period of five years, the savings could be $4.28 billion per year. (11)

Benefits of Virtual Care

It is clear that widespread deployment of telehealth during COVID and beyond has many potential benefits, including: greater access to doctors for populations in areas where medical clinicians are not always available; reductions in the numbers of people who are forced to live in institutional settings and can now remain at home using continuous monitoring; access to care with the press of a button; emergency medical care in trauma cases that occur far away from hospitals, and access to specialists in far-away cities. Just as e-Patients have aggressively pursued their quest for healthcare information online to address their individual illnesses and conditions, e-Patients must actively pursue telehealth applications and home monitoring services when and where they are needed.

Key Points

1.   Digital technology by way of virtual care to replace so many face-to-face encounters is now available to provide expert clinician care to pa-

tients when distance separates them.

2. Telehealth depends upon a robust infrastructure that includes the internet, video, high-speed high-bandwidth telecommunications, computer-based diagnostics, and various medical devices that enable doctors to virtually examine, treat, and interact with patients.

3. Home monitoring is critical to the care of those individuals who have chronic disease or a disability.

4. There are a variety of medical devices that can send vital statistics and other information to your doctors or nurses, so you are continuously monitored. This will save you from unnecessary visits to your clinician or to emergency rooms.

5. Virtual health enables your access to consults and second opinions with key experts from any major medical facility.

6. Telerehabilitation provides you with virtual physical therapy, hearing and speech therapy, occupational therapy, mental health services, and hospice services right in your home.

7. Ongoing maintenance programs such as the monitoring of asthmatic children, providing remote eye exams, and administering telerehabilitation and mental health services reduce the need for emergency medical procedures and save time, money, and the anxiety of having to transport individuals to a care facility.

8. In an emergency, telehealth can mean the difference between life and death.

9. e-Patients, when considering a hospital admission, might want to find out if the hospital they choose has an intensive care physician on staff or an eICU® program. The eICU® makes it possible for every intensive care unit in a hospital to have a physician who specializes in intensive care medicine checking patients at all times.

10. Private medical institutions, the states, and the federal government have to work together to vastly expand the telehealth infrastructure in the U.S. so that is more ubiquitous and available

Notes

1. Arizona Telemedicine Program, "A Life Saved through Tele-trauma Service." Arizona Telemedicine Newsletter p.4: http://www.federal-telemedicine.com

2. https://www.washingtonpost.com/national/out-here-its-just-me/2019/09/28/fa1df9b6-deef-11e9-be96-6adb81821e90_story.html

3. Robert Cox, MD, Medical Director at Hays Medical Center in Hays, Kansas, in an interview with the author, September 2009.

4. Lisa Vetter, Telemedicine Specialist, St. Alexius Medical Center and Telecare Network of North Dakota, in an interview with the author, October 2009.

5. Nina Antoniotti, Marshfield Clinic, Telehealth Network, Marshfield, Wisconsin, in an interview with the author, February 2010.

6. Edward D. Lemaire, Yvon Boudrias, Gayle Greene, "Low-Bandwidth internet-Based Videoconferencing for Physical Rehabilitation Consultations," Telemedicine, Telecare, Royal Society of Medicine Press, 7(2001):82-89, doi:10.1258/1357633011936200.

7. Statistics from National Institute of Mental Health http://www.apps-healthstatistics/healthstatistics/indexhelath.htm

8. Resources and Services Administration, anecdote from case studies on telehealth, Consultations Using Telemedicine, p5.

9. http://telemedicine/grants/success.html

10. Dr. Robert Cox Medical Director at Hays Medical Center, Hays, Kansas, in an interview with the author.

11. E. Pan, C. Cusack, J. Hook, A. Vincent, D.C. Kel, D.W. Bates, and B. Middleton, "The Value of Provider-to-Provider Telehealth Technologies" Telemedicine Journal E Health 14, 5(2008):446–53.

CHAPTER SEVEN

Patient-Centered Care

In spite of their growing independence, patients still expect to have a close, trusting, and personal relationship with their primary care physicians. They also want to be partners with their doctors.

Lawrence, David, MD, Chairman Emeritus, Kaiser Permanente. From Chaos to Care, Ad Capo Press, 2002, p.14.

A new model of care is evolving—patient-centered care—that creates a healthcare team consisting of you, your physician, possibly a physician's assistant or nurse practitioner, your family, and other healthcare professionals such as social workers and therapists. This team works together with you to ensure that you are receiving the coordinated care appropriate for your particular situation. In past generations, your care was based on your physician's evaluation of your health issues and the treatment recommended. This sea change in the care model now places the responsibility for your healthcare directly with you, in partnership with your providers. The IOM defines patient-centered care as "providing care that is respectful of and responsive to individual patient preferences, needs, and values and ensuring that patient values guide all clinical decisions." Further the IOM states that "twenty-first century medicine should be care based on continuous health relationships. Patients should receive care whenever they need it, and in many ways, not just face-to-face visits." Patient-centered care pre-supposes several things:

- That you and your family are encouraged to participate in your care.
- That you share in decision making leading to specific actions and treatment.
- That you direct your care, with the advice of your providers, when you are healthy and have a family advocate or friend do so when you are too sick.
- That you get all of the care that you need, within the boundaries of what your health insurance allows.
- That you have choices in determining who your providers will be based on performance data that is available to you.
- That you determine how your healthcare dollars are spent by becoming a better healthcare shopper who is able to find and evaluate purchases and understand the consequences of your choices.
- That your healthcare professionals communicate with you and your family and share complete and unbiased information.
- That the information you receive is timely, complete, and accurate and presented in a way that will help you effectively participate in care decisions.
- That all forms of communication—email, online technologies such as patient portals and discussion forums, clinical information systems, smartphones, and telemonitoring systems—are used in your interactions with your providers.
- That you are granted full access to care, not just via face-to-face and phone consultation, but via email and e-visits during off-hours.
- That there is good communication among physicians, nurses, and other health professionals taking care of you so that medical errors are reduced and duplication of tests and procedures is eliminated. (1)

The patient-centered care model requires a major commitment from providers and from you the e-Patient. It considers your cultural traditions, personal preferences, and values; your family situation; and lifestyles. It places responsibility for important aspects of self-care, such as blood-sugar or blood-pressure monitoring, in your hands. It includes the tools and support needed to fulfill those tasks. Multiple studies indicate that you are more likely to take your pills, eliminate bad food from your diet, and show up for appointments when you are encouraged to make your own choices. Patient-centered care is based upon the premise that there is a provider for every patient—a gatekeeper who coordinates and ensures continuity of care, even under the most extreme circumstances.

The Lancet, which is a collaborative, international team of over 200 professional editors, and publication experts who also follow the trends and major events in healthcare, digital health, cost of care, climate and health include a team of technical experts who publish several renowned journals. In a study entitled: Finance in healthcare and how to put people at the center. Their findings, published early in the twenty-first century, (April 4, 2022) revealed the following:

Primary healthcare (PHC) is widely recognized as a key component of all high-performing health systems and is an essential foundation of universal health coverage. However, in many places worldwide, PHC does not meet the needs of the people who should be at its center. Public funding is insufficient, access remains inequitable, and patients often have to pay out of pocket for services. Establishing the right financing arrangements is one crucially important way to support the development of people-centered PHC in most low-income and middle-income countries (LMICs), PHC is not delivering on the promises of these declarations. In many places across the globe, PHC does not meet the needs of the people—including both users and providers—who should be at its center. Public funding for PHC is insufficient, access to PHC services remains inequitable, and patients often have to pay out of pocket to use them. A vicious cycle has undermined PHC: underfunded services are unreliable, of poor quality, and not accountable to users. Therefore, many people bypass primary healthcare facilities to seek out higher-level specialist care. This action deprives PHC of funding, and the lack of resources further exacerbates the problems that have driven patients elsewhere. (2)

Your Medical Team

As we engage in the second decade of the twenty-first century, there are nearly fifty thousand fewer primary care doctors in practice than are needed to support the demand for basic medical care in the United States, particularly with legislation that mandates that every American citizen must have healthcare coverage. This was particularly evident during the pandemic. It means that nearly 20% of Americans (fifty-six million people) have inadequate or no access to primary care. Unfortunately, medical school graduates continue to choose specialty practice over primary care because the pay is better, the workday is shorter, and the liabilities are fewer. It is your primary care provider, however, who is charged with making sure that your basic care needs are covered, including annual physical exams, vaccinations, appropriate labs, and tests. Your primary care provider is also trained to inform you about good health practices, disease prevention, health maintenance, and treatment of your chronic and acute illnesses. They advise and educate you and coordinate care with all of your specialists, nurses, social workers, and therapists. They provide you with referrals and prescription refills and help you seek a second opinion when appropriate. This is basic healthcare, and it should be based on a trusted relationship. With the extreme shortage of primary-care physicians and little hope that this problem will be resolved in the near future, many institutions offer solutions. There are some solutions. HMOs, and family and internal medicine practices have turned to physician assistants, or nurse practitioners to help fill the gap.

A physician assistant is a healthcare professional licensed to practice medicine with the supervision of a licensed physician. The main focus of a physician assistant is prevention, maintenance of treatment for chronic conditions, education, and collaboration with patients to maximize care efficiency. Physician assistants conduct exams, diagnose and treat illnesses, order and interpret tests, counsel on preventive care, assist in surgery, and write prescriptions. Their education encompasses many of the same science courses and laboratory instruction taught in medical school. The training requires two to three years beyond undergraduate education. A physician assistant must also pass a National Certifying Exam (PANCE) before receiving a license to practice. They are required to earn and log continuing education credits and renew their license every two years. Every six years, they must take an exam for recertification.

Some primary care practices use nurse practitioners to provide basic care services. There are also independently practicing nurse practitioners. These individuals also must have a license to practice, which qualifies them to see patients, give exams, make diagnoses, collaborate with physicians, make referrals, and diagnose and treat acute illness, injuries, and infections. They also write prescriptions, counsel patients on health behaviors, manage chronic conditions, and make referrals to various therapists. Most nurse practitioners specialize in a specific area of medicine (e.g., family medicine, pediatrics, obstetrics and gynecology, or diabetes).

Using physician assistants and nurse practitioners is a cost-effective way to maintain effective individualized care, and you can rest assured that they are well qualified to handle your basic medical care and will have time to thoroughly address your concerns. These professionals use the tools of technology, including digital health records, email, smartphones, patient portals, e-visits, and telemedicine, and they communicate with your primary care doctors and specialists. What is important is that you have a central place where you are receiving your basic medical care.

The Patient-Centered Medical Home

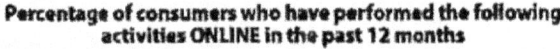

Percentage of consumers who have performed the following activities ONLINE in the past 12 months

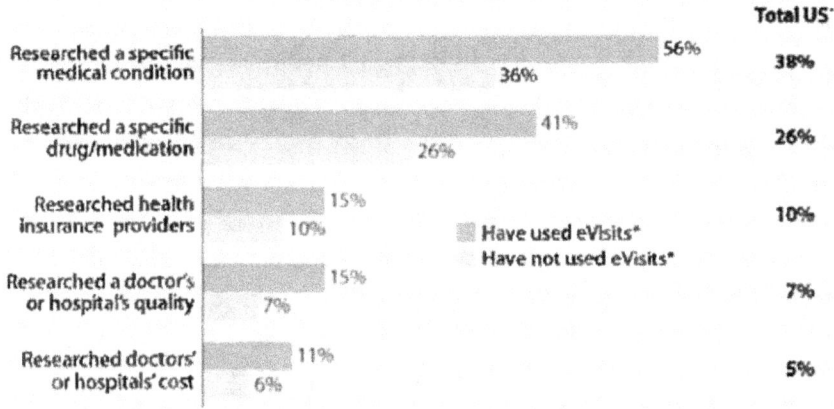

*Base: US consumers whose primary doctor offers eVisits
'Base: US consumers

With available technology, a new model of care has evolved: the patient-centered medical home (PCMH), where you choose a primary care provider who could be a physician, physician's assistant, or nurse practitioner. Access to a specialist is available when needed. In the PCMH, you are at the center of the healthcare experience, supported by your team of caregivers and by the technology. This includes a secure portal, email, and an electronic health record that contains your lab tests and your medical history. Your provider team helps you find links to websites that explain various conditions and enable you to engage in online discussions. You also have the convenience of flexible scheduling, which includes e-visits and e-prescribing. Patients who have a medical home stay healthier. This is because the staff at your medical home is familiar with your medical history and can help with preventative care and managing chronic conditions such as high blood pressure, diabetes, and asthma. Reimbursement in a PCMH model includes not only the usual fee for service payment but a retainer fee system, typically paid by your health insurer, that pays for services associated with coordination of care between your medical home practice, consultants, therapists, and community resources.

The Patient-Centered Primary Care Collaborative is a coalition of employers, consumer groups, health plans, and healthcare professionals that promotes the concept of the PCMH. The collaborative conducted several pilot projects with results indicating that patients who had a PCMH had far fewer emergency room visits and hospitalizations. Additionally, among patients with chronic conditions, there was as much as a 40% decrease in hospitalizations and nearly 30% reduction in ER visits. These patients who had a medical home appeared to stay healthier and experienced significant healthcare cost savings (3)

In 2007, Group Health of Seattle decided to experiment with a PCMH model of care. Group Health reported the following: During the two years studied, the team's patients had 29% fewer ER visits and 6% fewer hospitalizations compared with other Group Health clinic patients. There were start-up costs—sixteen dollars per patient per year—and it took a couple of years to provide the bulk of the savings. However, for every dollar invested in the system, Group Health saved $1.50 by keeping patients out of the ER and the hospital. Medical home patients reported better care experiences.

The strategy is now being expanded to all twenty-six of Group Health's Washington-state medical centers, covering more than four hundred thousand patients. Home monitoring tools such as weight and blood pressure measurement devices that transmit readings via the internet both to children of elderly home-bound parents and to a healthcare provider enable not only the relatively healthy patient to be a part of the medical home but also provide a way for the elderly to remain in their own homes rather than go to assisted living and nursing home facilities, yet they still have the assurance of being cared for and safe.

The American Academy of Pediatrics (AAP) in a policy statement advocates that every child deserves a medical home where care is accessible, continuous, comprehensive, patient—and family-centered, coordinated, compassionate, and culturally effective. The AAP suggests the following characteristics for a medical home for children:

- Family-centered care based upon a collaborative working partnership with families, respecting their diversity and recognizing that they are the constant in a child's life.
- A community-based system with a coordinated network of services that promote the healthy development of children and includes on-going primary care as well as access to a broad range of specialty and ancillary services.
- Optimization of the lifelong health that continues uninterrupted as the individual moves through the system from adolescent to adulthood. (4)

Pilot PCMH programs conducted by Medicaid demonstrate that coordinated patient-centered care using the medical home concept contributes real value to care and reduces cost. A Medicaid study documented a savings of $3.5 million over three years for asthma management and $2.1 million savings for a diabetes care initiative. These savings accrued from elimination of wasteful duplication of services, tests, and procedures and from close control of chronic illnesses.

While the medical home is not a silver bullet that will solve all the problems of twenty-first century healthcare, it has the potential to transform healthcare by blending personalized care with team-based coordinated care that deploys technology to help you understand and work through complex healthcare issues. You have to seek your own medical home, either with the right primary care practice or through a clinic or health center. You also have to do the work to ensure that you supply all of the necessary elements for the best patient-centered care available. To do that you must:

- Put together an accurate and current medical history—a personal health record.
- Maintain complete and honest medication lists for prescription and over-the-counter drugs.
- Follow individualized care plans outlined by your healthcare provider, including medication adherence.
- Maintain an accurate record of allergic reactions to food and medicines.
- Understand and own your healthcare costs.

- Provide feedback to providers, such as blood sugar and blood pressure daily records.
- Engage in email communications and e-visits instead of always heading to the physician's office or the emergency room or depending upon telephone calls.
- Go online to read test and procedure reports, to do scheduling and referrals, and to educate yourself about various health concerns,
- Work with the medical team to engage in preventive care.
- Do the necessary research on the internet to understand personal and public health issues.
- Advocate for your own health needs or those of family members; do not wait for the physician to identify those needs.
- Provide a personal health record that ideally links to your physician's electronic medical record.

Patient-Centered Care Alternatives

Retail Clinics

In the ideal healthcare environment, you would always go to your primary care office for care. But it is not always possible. Retail clinics, which sprouted in the early days of the twenty-first century, are a good alternative for care when options (other than the emergency room) are not available. These clinics mainly take care of basic health needs. They are set up to provide care on a walk-in basis (no appointment required) during regular hours as well as evening and weekend hours. With the digital tools now available, a summary of everything that occurs during your visit to a retail clinic can be transmitted to your primary care physician. Minor health issues that used to send you to the hospital ER, where the wait is interminable and the cost prohibitive, can be resolved in a retail clinic.

When you visit a retail clinic, you are generally seen by a licensed nurse practitioner who is more than qualified to handle your nonemergency health concerns, such as colds, coughs, ear aches, cuts and bruises, stomach complaints, flu-related symptoms, and strep throat. They can also administer standard flu shots and other injections. Many clinics also offer routine physical exams. Some have expanded their scope of service to include management of chronic conditions or setting of broken bones. Others are conducting smoking cessation programs and counseling related to HIV and STDs. Payers including Aetna, CIGNA, and United Healthcare, and some of the Blues now reimburse for retail clinic visits. Some clinics have a loose affiliation with a local hospital. They are subject to oversight by governmental health agencies, who see that they maintain a level of quality and safety in their practice. Unlike a visit to your doctor, pricing for various services at a retail clinic are posted on a menu board, and if your payer approves the use of the clinic, you just have to make the copayment required. If you need to have a prescription filled, there is usually a pharmacy right there. Retail clinics are also a great option for people who are traveling, fall ill, and can't get in touch with their regular healthcare provider. If you go to a retail clinic, you must be sure to bring all necessary health information with you, including your medications, health history, and contact information for your regular provider so that you will receive appropriate treatment.

A study by Rand Health showed that a large percentage of people who use retail clinics do not typically have a primary care provider. The population at these clinics tends to be younger patients, minority families, and families with children, often without health insurance. There are very few patients age sixty-five and older. Convenience and low-cost services, short wait times, transparent pricing, and dissatisfaction with existing primary care are the primary reasons study respondents cited for visiting a retail clinic. Many of these patients pay for the retail clinic services out of pocket, and nearly 25% of patients who use retail clinics do not have health insurance. The study reported that overall satisfaction with quality of care at the retail clinic is high. In most instances, the retail clinic will refer a patient who needs more complex oversight to a primary care practice or to a hospital. (5)

Community Health Centers

Community health centers (CHCs) and free clinics evolved from President Lyndon Johnson's War on Poverty program, which, in the middle of the 1960s, allocated federal funding for millions of US citizens without health insurance who had great difficulty finding providers to accept them for care. Today, funding for CHCs is tied to several federal and state programs designed to provide health services to underserved populations in both rural and urban areas. Medicaid is the largest source of revenue for CHCs, followed by federal grants.

The ambulatory care services offered by these health centers reflect the diverse needs of the population they serve. Generally, there is a high demand for gynecologic, family practice, and pediatric services. At the bare minimum, all of the centers offer wellness checkups, immunizations, basic tests, and treatment when you are sick. Many also offer dental care, prescription drugs, and mental health and substance abuse care. The populations served include individuals receiving unemployment benefits, individuals employed in seasonal or part-time work, and those who are experiencing financial hardship and do not have health insurance coverage. Many health centers also offer an array of public services such as translation, transportation, outreach, eligibility assistance, and health education. Community health centers include a large population of young patients who come with a high incidence of chronic conditions like diabetes, asthma, and hypertension in disproportionate numbers to the general population. The centers are often staffed by local physicians, some of whom donate their time. Nurses, nurse practitioners, and lab technicians also staff these facilities. A dispensary is usually located on site where prescriptions are filled. (Narcotic drugs are not available.)

A key provision of the Health Reform Act of 2009 expands the national network of community health centers. The reform legislation increased CHC funding by $11 billion dollars between 2010 and 2015. The legislation includes payment incentives to encourage high quality care and expansion of information technology infrastructure so that these centers can serve as patient-centered medical homes. Multiple studies have documented health centers' efficacy in reducing the delivery of low-birth-weight babies and reducing the number of hospitalizations for patients with chronic conditions. Only one-third of health centers in the United States are accredited by the Joint Commission on Accreditation of Healthcare Organizations (JCAHO), a private not-for-profit organization that operates accreditation programs for hospitals and other healthcare organizations throughout the United States. JCAHO sets the standards of care to which healthcare organizations aspire. A majority of state governments have come to recognize JCAHO accreditation as a condition of licensure and therefore permit the receipt of Medicaid reimbursement at those locations. The statute that created CHCs includes the following criteria that must be followed for each center that is established:

- **They must be located in or serve a high-needs community.** These medically underserved areas are defined as having a high percentage of people living in poverty, areas with few primary care physicians, higher than average infant mortality rates and high percentages of the elderly.

- **They must provide healthcare to all, regardless of ability to pay.** All community health centers must commit to providing services for everyone, with fees based on a standard a sliding fee schedule that adjusts charges for care according to income.

- **They must provide comprehensive healthcare services.** All community health centers also must offer a broad range of "enabling" services to support the delivery of consistent, affordable healthcare.

- **They must be governed by a community board.** All community health center boards must be comprised of a majority (at least 51%) of health center patients who have the authority to oversee the operations of the center. These powers include approving budgets, hiring and firing chief executives, and establishing general policies.

These centers now have a forty-five-year history of providing care in underserved communities for everyone, regardless of their ability to pay. The major issues are having enough providers to treat all the patients who come to them and educating patients to think about their health before there is a crisis. The Sioux Falls Community Health Center includes two satellite clinics located in the public schools, where they provide general health services, flu shots, and other inoculations. Although they may not meet all of the "official" requirements of a patient-centered medical home, to the over twelve thousand individuals that they serve, the Sioux Falls Community Health Center is their medical home, and they come back again and again for medical care and services that go above and beyond basic medical needs. (6)

A Commonwealth Fund survey of community health centers found that the more medical home characteristics a center possessed, the more likely it is to report better communication and coordination with specialty care providers and local hospitals. The study further indicated that 40% of health centers use electronic medical records, which is nearly the same rate as practitioners throughout the country. Fifty-seven percent of centers reported they electronically access patients' laboratory tests results on a routine basis, and 45% routinely order laboratory tests electronically. The report found that enhancing a center's use of health IT is related to better coordinated patient care.

For nearly thirty years, a community health center located in downtown

Sioux Falls, South Dakota, has seen thousands of patients from a variety of backgrounds and nationalities, including underserved individuals who have lived in the area for many years and an influx of refugees from countries throughout the world who recently migrated to this region as a part of the refugee resettlement program sponsored by the US Department of Health and Human Services. The Sioux Falls Community Health Center is a large-scale operation that includes family practice clinics, mental health services, pediatric services, and a Ryan White early intervention program for HIV and AIDS patients, as well as dental and vision services. Interpreters are available for patients who have little or no English skills. The staff also includes nurse case managers and social workers who counsel patients and educate them in health management and prevention, and a volunteer cardiologist and dermatologist. The Sioux Falls Community Health Center includes two satellite clinics located in the public schools, where they provide general health services, flu shots, and other inoculations. Although they may not meet all of the "official" requirements of a patient-centered medical home, to the over twelve thousand individuals that they serve, the Sioux Falls Community Health Center is their medical home, and they come back again and again for medical care and services that go above and beyond basic medical needs.

According to Katie Wick, Clinical Services Manager, the Sioux Falls Community Health Center in 2009 served over 12,200 patients who had over 40,000 visits, of which 28% were pediatric visits. The number one diagnosis that the clinic addresses is depression, for which they have counselors and a psychiatrist available. There is also a high percentage of treatment for diabetes and cardiovascular disease. The dental clinic is staffed by three dentists. In partnership with the local Lions Club, they also offer eye exams. They even have vans that go out into the streets of Sioux Falls to pick up homeless people and drive them to the center for treatment. With or without health insurance, everyone who comes to the Sioux Falls Community Health Center is treated. Over 50% of patients do not have health insurance or are underinsured; the other 50% have Medicaid, Medicare, or a private health plan. Patients have to pay a minimum of fifteen dollars for a medical visit and thirty-five dollars for a dental visit. A sliding fee structure is offset by federal grants. Many patients are homeless or in transition. (7)

For more than forty-five years, community health centers have delivered comprehensive, high quality preventive and primary healthcare to patients regardless of their ability to pay. Once a small dot on the healthcare landscape, CHCs have become the essential primary care medical home for millions of Americans including some of the nation's most vulnerable populations in medically underserved urban and rural communities. In 2018, health centers served over 28 million patients – a 31% increase in only five years. This expansion resulted from major investments by Medicaid expansion, which dramatically expanded patient coverage, and by the Community Health Center Fund, which provided direct investment in health center growth

Even more community health centers are vitally needed in this country as the pandemic so starkly indicated. CHCs are often are the only place where people in the underserved communities could receive a test for COVID, a vaccination, or medical assistance They are often the only place where women can access gynecological and obstetrics services. (6)

E-visits

Online health consultations, or e-visits, enable you and your physicians to connect on the internet using a secure portal site to address nonemergency issues and important questions you may have. They help maintain a continuum of care. Messaging software enables you to log into a secure website and fill out templates and logic-based questions, which are sent to a physician who will answer your communication and engage in a dialogue with you asynchronously. You may send a request for an e-visit, and it can take twenty-four hours, or as much as seventy-two hours on a weekend, for a response.

The benefit of an e-visit is that it saves you time and money, since you do not have to travel to see your doctor in person. Make no mistake about it—the e-visit does not replace your face-to-face visit with your doctor for more important and comprehensive medical issues. E-visits should be used for only minor ailments such as colds and sore throats, rashes, or bruises. You might also request an e-visit in instances where you can send a digital photo ahead for discussion or for a periodic review of a chronic condition. It is also an efficient way to have a general follow-up consultation with your doctor in between visits. A growing number of health plans, including Aetna and Cigna, have begun to pay doctors for online visits with their patients.

E-visits should be based on personal interaction with your provider and should occur only when you and your physician have a preexisting relationship that includes a face-to-face encounter within the current year. Unfortunately, many patients use this technology for online medical consultations at one of the hundreds of websites that offer free or nominally priced medical consults with doctors you have never met. These consultations are priced at between twenty-five and forty dollars—the approximate amount that you might typically be charged for co-pay for a regular office visit. The medical professionals who are talking to you at these sites have no reference points regarding your medical history. Although you can get quick answers, the individuals you are talking with could misinterpret your problem or make erroneous conclusions because they are not familiar with your specific case. People surfing the internet for health information admittedly often do not check the source and date of the health information they are reading. Engaging in an e-visit with a physician who does not have access to your health record and whom you have never met is not wise. The American Academy of Family Physicians states on their website that they support enhanced-access physician-patient interactions, including virtual/electronic visits or "e-visits," that occur over safe, secure, online communication systems.

Their guidelines for e-visits include the following:

- E-visits should be used only when you have an established relationship with your provider.
- The patient initiates the process and agrees to e-visit service terms, privacy policy, and charge for receiving asynchronous care from a physician or other qualified health professional.
- Electronic communication occurs over a HIPAA-compliant online connection.
- An e-visit includes the total interchange of online inquiries and other communications associated with this single patient encounter.
- The physician appropriately documents the e-visits, including all pertinent communication related to the encounter, in the patient's medical/health record.
- The physician or other qualified health professional has a defined period of time within which responses to an e-visit request are completed.
- E-visits should be a payable physician service. (9)

The Henry Ford Medical Group of the Henry Ford Hospital is a part of the world-renowned Henry Ford Health System located on West Grand

Boulevard in the New Center area of Detroit. It is one of the nation's largest group practices, with over one thousand physicians and researchers in more than forty specialties. Henry Ford offers e-visits to their patients. To initiate an e-visit, patients go online to their portal (MyHealth) at www.henryford.com and login to their secure page. They have two options:

1. Describe their nonurgent health and medical issues to their primary care provider. Based upon their question, a series of queries will be posted to the patient; subsequent questions are based on their responses. The branching logic in the interviews provides physicians with the information they need to assess the patient.
2. Describe their nonhealthy issues in a preset series of questions around the following chronic conditions: diabetes, hypertension, depression, asthma, and coronary artery disease.

Henry Ford physicians can respond to both types of visits in a number of ways: by ordering lab tests, changing or ordering new prescriptions, instructing the patient to take action at home, or advising him or her to come into the office. Each e-visit costs twenty dollars. Patients are not charged if their issue is not resolved and an office visit is necessary.

Initially, the program designers thought that people were going to use e-visits for low-acuity health problems; however, their experience with the e-visits proved otherwise. Physicians in the system found e-visits to be most helpful for those patients suffering from chronic conditions because it enabled primary care physicians to make more frequent adjustments to their patient's treatment regimen and achieve better clinical goals between office visits. Henry Ford also thought young people would flock to this program because they tend to be online more, but that did not prove to be the case either. Instead, they found the average age of the e-visit user was forty-four to fifty-one years and a person who suffers from a chronic condition. Henry Ford also realized that e-visits have a number of benefits, including enhancing patient and physician communication. (10)

A CMS ruling established in 2009 allowed physicians to bill Medicare for electronic consultations (e-visits) following inpatient visits. It also adds new codes specifically for telemedicine consultations for healthcare providers who are consulted by a patient's physician but are not available for in-person consultations. These e-visits, performed in real time with interactive communications systems, are for the purpose of monitoring a patient's progress, recommending care-management changes, or providing a new plan of care. Once CMS makes the e-visit a standard, reimbursed part of payment to physicians, most payers will follow.

There are several leading digital care online services including: Amwell, Doctor on Demand, Iclinicq Teledoc, Zocdoc, and more, that provide an option for patients to access healthcare assistance 24/7 for an out-of-pocket fee. These options are especially useful on weekends, holidays, evenings and other off-hours, when your primary care doctors are often not available. Many employers, as well as individuals, now contract with online health services for a monthly fee, to supplement their regular care or to substitute online care for regular in-person care, in situations when in-person care is only available by going to an emergency room. Using a phone app or tablet and two-way video services, online options provide you with interactive choices to basic clinical care, pediatrics, behavioral health consults, nutrition and dermatology, as well as chronic care management. When you connect to one of these websites, you fill out a health history which creates a digital record. With the patient's permission, a copy of the visit notes is sent to your regular primary care provider so your care is coordinated and safe. You need to ensure that your regular provider, as well as the online physician, always has full information at the point of care, and that visit information is communicated to a primary care provider after you complete the online visit.

A Forrester Research study that looked at interactions between patients and care professional that occur online and what users are generally doing online. The study found that over 50% of individuals who engage in e-visits also have researched a specific medical condition online, and over 40% have researched a specific medication. This mail survey, involved more than five thousand individuals who live in North America. (11)

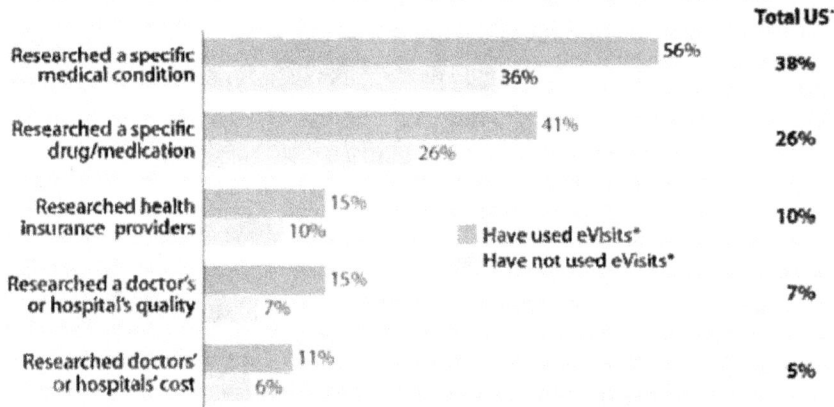

Percentage of consumers who have performed the following activities ONLINE in the past 12 months

		Total US·
Researched a specific medical condition	56% / 36%	38%
Researched a specific drug/medication	41% / 26%	26%
Researched health insurance providers	15% / 10%	10%
Researched a doctor's or hospital's quality	15% / 7%	7%
Researched doctors' or hospitals' cost	11% / 6%	5%

Have used eVisits*
Have not used eVisits*

*Base: US consumers whose primary doctor offers eVisits
'Base: US consumers

House Calls

In an era when many patients complain about paying too much money for too little time with their doctors, Dr. Susan Rutten Wasson, who is trained as an internist and pediatrician, has a different approach that is as close to patient-centered care as you can get. Dr. Wasson, who has been in practice for ten years, sees patients by appointment in her small- town office in Osakis, Minnesota. She allows her patients same-day scheduling for office visits and spends at least a half hour with each patient, for which she charges a fifty-dollar cash fee out of pocket. Dr. Wasson accepts no insurance and thus is not bogged down with billing third-party payers. Prior to opening her office in Osakis, Dr. Wasson's practice consisted primarily of house calls. She still makes some house calls today to help patients who have difficulty getting to her office.

"You learn a lot about people when you see where and how they live," she says. Dr. Wasson does not schedule regular visits; her patients come to see her when they need to, although she does see individuals with chronic conditions at least once a year. In a typical visit, Dr. Wasson will check blood pressure, weight, and other vitals and talk with her patients in detail about what is bothering them. She sends her patients to a local clinic to have blood drawn and refers her seriously ill patients to one of the local hospitals or to the major medical centers in either the Twin Cities; Fargo, North Dakota; or Sioux Falls, South Dakota; or she sends them to the Mayo Clinic. Most of the individuals she serves are what she calls "good country people" who make their living by farming or trucking. Many of them have never used a computer, cell phone, or email and have no access to technology. English is a second language for many patients, and computers are not an option.

Dr. Wasson is a firm believer that people must take responsibility for their own health, but she clearly guides them in making decisions. She follows the tenets she learned in medical school to run through a list of possibilities and to examine her patients carefully and narrow down the possibilities. "Too many tests," she says, "get ordered in a shotgun approach." She believes in sending her patients for tests (e.g., CAT scans, MRIs, X-rays, or extensive blood work) only when the visual diagnosis is unclear. She is convinced that people in general have no idea of what these procedures cost until they see a bill. Patients hear about Dr. Wasson from the local pharmacist and from other patients. Many come to her for a second opinion. She is always busy.

Several physicians and physician groups are once again making house calls to home-bound patients who cannot get to the doctor's office. This represents the very definition of patient-centered care. Today's physicians who do home visits can not only do a thorough physical exam, but can perform EKG testing, draw blood, and even do an X-ray using the latest digital machinery, which is portable and efficient. Furthermore, with electronic records, they can record all of their test results and observations right on a laptop for communication to various entities, including hospitals, payers, and other doctors or specialists you might see.

New York City's Presbyterian/Weill Cornell Medical Center has sponsored a program called "House Calls" since 1977 to meet the needs of a portion of the elderly home-bound city population who can no longer navigate their way to their doctor. The program serves households in neighborhoods around the hospital and enables the doctors in the program to spend quality time with their patients and get to know them well. (12)

Microsoft brought back the old-fashioned doctor's house call for employees in a bid to slash costs, improve employee health, and contain potential pandemics. Microsoft is one of the few employers, and by far the largest, offering such personalized health services to workers. Their house call program, called Mobile Medicine, got its start in 2006 after a massive analysis of the company's healthcare data revealed that their employees, like others, were using the emergency room mostly for non-life-threatening, problems such as ear infections, skeletal bruises, and the flu. Such trips to the emergency room average $1,500–$2,000 a visit, according to the American Academy of Home Care Physicians. As a result, Microsoft's policy of having family doctors show up at their employee's bedsides, makes economic sense. Microsoft's data claims that they have reduced healthcare costs by approximately 35% with the house call program. Additionally, the company claims that the strategy gives primary care doctors more responsibility to manage patients' treatments and keep employees healthier in the long term. (13)

Why Patient-Centered Care?

As an e-Patient, your goal has to be focused on ensuring that you are receiving the best care that current technology and e-trained medical professionals have to offer you, within a reasonable healthcare budget. This is why finding a medical home where your basic health needs can be met is important. It is also why, when you are stymied by the system, you can turn to a retail clinic, a community health center or even an online provider for care. It is also why you want to advocate access to a secure web portal that enables you to engage in web visits and access your health information so you are in control of managing your care. At the end of the day, patient-centered care and the medical home will bring about reforms both in terms of financial management and in quality of care. It will encourage the spread of participatory medicine that is critical to resolving some of the issues regarding fair and equitable healthcare for all citizens.

What Patients Want

% Responding Somewhat or Strongly Agree

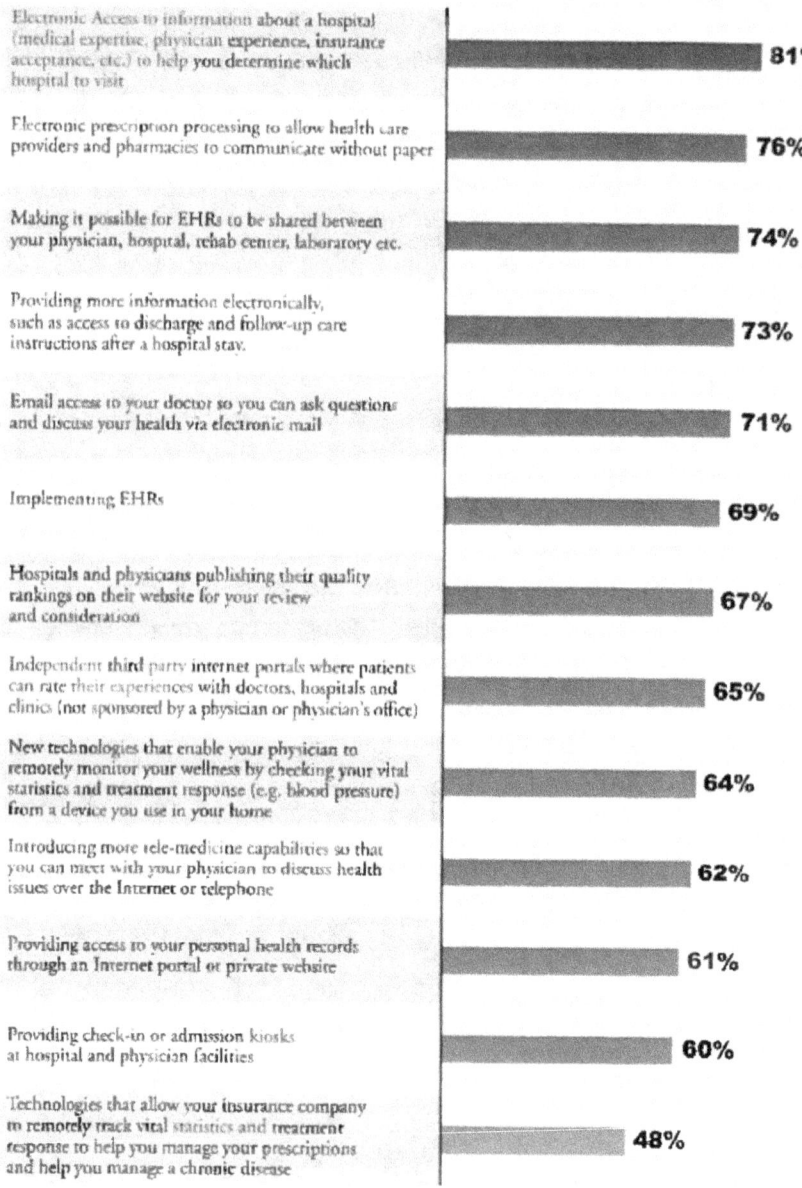

Electronic Access to information about a hospital (medical expertise, physician experience, insurance acceptance, etc.) to help you determine which hospital to visit — **81%**

Electronic prescription processing to allow health care providers and pharmacies to communicate without paper — **76%**

Making it possible for EHRs to be shared between your physician, hospital, rehab center, laboratory etc. — **74%**

Providing more information electronically, such as access to discharge and follow-up care instructions after a hospital stay. — **73%**

Email access to your doctor so you can ask questions and discuss your health via electronic mail — **71%**

Implementing EHRs — **69%**

Hospitals and physicians publishing their quality rankings on their website for your review and consideration — **67%**

Independent third party internet portals where patients can rate their experiences with doctors, hospitals and clinics (not sponsored by a physician or physician's office) — **65%**

New technologies that enable your physician to remotely monitor your wellness by checking your vital statistics and treatment response (e.g. blood pressure) from a device you use in your home — **64%**

Introducing more tele-medicine capabilities so that you can meet with your physician to discuss health issues over the Internet or telephone — **62%**

Providing access to your personal health records through an Internet portal or private website — **61%**

Providing check-in or admission kiosks at hospital and physician facilities — **60%**

Technologies that allow your insurance company to remotely track vital statistics and treatment response to help you manage your prescriptions and help you manage a chronic disease — **48%**

Key Points

1. A patient-centered model of care provides a care team including the individual, the healthcare professionals, the family, and all others who may be involved with the individual's care, including social workers and therapists. These individuals share responsibility for the patient collaboratively in a patient-centered medical home that ensures that the patient will experience the safe, high-quality, continuous care that you deserve.

2. Digital communication technology, including the internet, secure portals, email, e-visits, and smartphones, are critical components of the medical home.

3. Physician assistants and nurse practitioners fill the personnel gap created by a critical shortage of primary care physicians. They serve as the primary caretakers for individuals who are triaged to doctors when needed.

4. Retail clinics are spreading throughout the country and provide alternative care models for individuals who need immediate care when their medical home is not available.

5. Community health centers provide medical homes for the uninsured, the poor, and those who cannot access care via their workplace and health insurers.

6. E-visits, which are usually reimbursed are valuable to the patient for those non-urgent medical issues and the management of chronic conditions.

7. House calls are becoming more commonplace, especially in the inner city where elderly and disabled patients are not able to get to the doctor's office or a health center, and in rural locations where patients cannot travel the distances required to see the doctor.

Notes

1. National Academy of Sciences, Formulating New Rules to Redesign and Improve Care, Crossing the Quality Chasm, National Academy of Sciences, 2001, p. 61.

2. The Lancet Global Health Commission on financing primary health-care: Putting people at the center, April 4, 2022.

3. http://www.thelancet.com/commission/financing- primaryhealthcare.

4. Patient Centered Primary Care Collaborative, Patient Centered Medical Home: Building Evidence and Momentum: A Compilation of PCMH Pilot and Demonstration Projects, PCPCC, 2008, pp. 51–53, http://www.pcpcc.net.

5. Policy statement, The Medical Home. Pediatrics, 110 (2002): p.184–186, http://www.aap.org/health topics/health topics

6. Ateev Mehrotra, MD; Hagnsheng Liu, PhD; John L. Adams, PhD; et al., Comparing Costs and Quality of Care at Retail Clinics with That of Other Medical Settings for Three Common Illnesses, Rand Health, 2010.

7. Health Centers and the Affordable Care Act (hrsa.gov) Health Services and Administration Bulletin, Department of Health and human Services, 2014, http://www.hrsa.gov

8. Katie Wick, Clinical Services Manager, Sioux Falls HEALTH AF-FAIRS FOREFRONT

9. Keeping Community Health Centers Strong During the Coronavirus Pandemic is Essential to Public Health, Peter Shin, Rebecca Morris, Maria Casoni, Sara Rosenbaum, Alexander Somodevilla, Health Affairs , April 10, 2022.

10. American Academy of Family Physicians (AAFP) Guidelines, e-Visits, http://www.aafp.org/online/en/home/policy/policies/e/ evisits.html.

11. Pamela Landis, Director of Information Technology at Henry Ford Health Systems, interview with the author.

12. Which Consumers Are Healthcare eVisit Early Adopters? Elizabeth Boehm., Forrester Research, June 3, 2010.

13. New York-Presbyterian, House Calls Program at New York-Presbyterian/Weill Cornell Helps City's Elderly, New York-Presbyterian, Dec. 16, 2003, http://www.nyp.org/news/hospital/house-calls.html.

14. Michelle Conlin, Companies Eager to Cut HealthCare Costs May Do Well to Consider the Home Doctor Visits Microsoft Offers, Bloomberg BusinessWeek, November 5, 2009.

CHAPTER EIGHT
Web Resources

Advances in computer technology and the internet have changed the way America works, learns, and communicates. The internet has become an integral part of America's economic, political, and social life.

President William J. Clinton, President of the United States (1993–2001)

Healthcare Finds the Internet

The internet, which was the creation of scientists and US military specialists, now gives you instant, efficient communications that can transmit your information in nano-seconds. Built on a philosophy of openness and bound together by an endless infrastructure of digital connections, the internet allows you to ship the entire collected works of Shakespeare across the Worldwide Web in the blink of an eye.

Today, your personal healthcare information, which used to be the exclusive province of a healthcare professional, is available to you on the internet 24/7, if you choose to store it that way. With the spread of wireless technology, you can access that information not only from a desktop PC but from a wireless laptop, from a handheld device, and from your smartphone. You can also find important, relevant answers to your health questions, access independent medical diagnosis, contact health services, and connect with individuals who suffer from health issues similar to yours.

A young woman living in Oregon gave birth to a premature baby girl who had lung issues that could cause brain damage. The woman and her husband were devastated. Their doctors and nurses, although caring, competent, and sympathetic, were not able to give them a lot of hope or much information. In desperation, the woman and her husband turned to the internet, and they found web communities where they connected with other couples who had premature infants with similar conditions. Although the internet did not offer a magic formula for curing their baby, the web offered them hope as well as suggestions about new tests and drugs, which they were able to discuss with their medical providers, and take advantage of to find a remedy for their baby, who fortunately responded well and got better.

The Pew Research Center (Pew Internet and American Life Project) is an independent opinion research group that focuses on media and public policy. They conduct studies on the behavior of the American public and the internet. In April 2009 they released a definitive study, Social Life of Health Information, which looked at how the American public used the internet to find health information. The research was based on data from telephone interviews conducted by Princeton Survey Research Associates between November 19 and December 20, 2008, among a national sample of 2,253 adults. The results indicated that in 2008, 74% of American adults were online, and 61% of those adults look online for health information. This represents a significant change since the Pew internet research conducted in the spring of 2000, when 46% of American adults had access to the internet, and 25% looked online for health information. Other interesting results of this study revealed that:

- 52% of all online health inquiries are on behalf of someone other than the person typing in the search terms
- 41% of people looking for health information have read someone else's commentary or experience about health or medical issues on an online news group, website, or blog.
- 24% consulted rankings or reviews online of doctors or other providers.
- 24% consulted rankings or reviews online of hospitals or other medical facilities.
- 19% signed up to receive updates about health or medical issues. (1)

In 2011, Pew conducted another study, Health Topics, on the uses of the internet for gathering healthcare information. This study revealed that of 80% of internet users now look online for health information, making healthcare the third most popular online pursuit. This report is also based in part on a national telephone survey conducted by Princeton Survey Research Associates International that included 3,001 adults.

A profile of some of the topics that most interested these internet health information seekers in 2011 include:

- 66% look online for information about a specific disease or medical problem.
- 56% look online for information about a certain medical treatment or procedure.
- 44% look online for information about doctors or other health professionals.
- 36% look online for information about hospitals or other medical facilities
- 33% look online for information related to health insurance, including private insurance, Medicare, or Medicaid.

About six in ten say their internet searches have an impact on their own health or the way they care for someone else. Sixty percent say the information found online affected a decision about how to treat an illness or condition and 50% indicated that their online searches have changed their overall approach to maintaining their health or the health of someone they care for. (2)

Clearly the internet now attracts an increasing number of people who are seeking answers that they are not finding using other sources. On any given day, there are more people online seeking medical advice than actually visit health professionals. These individuals search for health information, research a diagnosis or prescription, prepare for surgery, find out how to best recover from an illness, or seek emotional support for a unique problem. Although the first and foremost source of information on your health should always be your primary care physician, there is a lot to be gained by supplementing your understanding of health issues by checking credible internet resources and connecting with others. Online support groups provide a particularly valuable asset. Finding individuals with the same or similar conditions who can help you understand all of your options could make a significant difference in your eventual outcome. Today, there is a website for virtually every specific disease or problem, where you can find detailed information that outlines treatment options and connects you with people who understand what you are going through. You merely have to type the name of the disease in your browser window and chances are, you will find a group to talk with.

During the COVID-19 pandemic, the internet was the most used source for health-related information and the uptick in the use of social media to stay abreast of what was happening and discuss common experiences became ubiquitous. Exploring the internet for new information about the pandemic became a daily occurrence in most households and the use of social media to connect with friends, family and strangers to discuss Covid issues was endemic.

Ken was diagnosed with fibromyalgia (FM) in 1996, following twelve years of searching for answers about why he had periods of numbness in his extremities and flu-like symptoms. After finally receiving the diagnosis, he found that his doctors had little to offer him by way of remedies or ideas on how to treat the disease. Ken and his wife, Donna, began searching the internet for information from their home in Regina, Saskatchewan. In 1996, there was not much available regarding FM. They found scattered bits and pieces of information, but nothing comprehensive. They decided to put together what they had collected and create a website that they called "Fibrohugs." Soon they began to get emails from people all over the world thanking them for developing the website. They set up message-board forums and later added a live chat room. The twenty to twenty-five people who gathered at Fibrohugs to chat on a regular basis began to feel like a little family. Over time, doctors recommended the site to their patients. By the end of 2009, the site had nearly twenty-seven thousand members, and over four hundred million individuals had visited Fibrohugs.

Five to ten new people join as members every day.

About three years ago, when a family member of Ken's developed cancer, Ken and Donna discovered there was a serious lack of one-on-one personal support available for cancer patients. They launched a sister site to Fibrohugs called Cancer Hugs. Ironically, four months into building the site, Donna was diagnosed with cancer. The Fibrohugs family suffered with them, encouraged them, and mourned with Ken when, sadly, Donna passed away. (3)

Information Access

Ben was diagnosed with prostate cancer. The treatment options for his particular form of cancer were thoroughly explained by his doctors, in whom he had confidence, but even with a second opinion consult, he was confused as to which way to turn. He had to decide among the choices: watch and wait (to see if the tumor grows larger), radiation therapy, radiation followed by surgery, surgery alone, or hormone treatment. To add to his confusion, there are different surgical approaches, some of which are more invasive with more potential aftereffects than others. His doctors could not tell him which option might be better. Ben turned to the internet and went to several cancer sites, joined chat rooms and discussion groups, and talked with people who had been in the same circumstances and who could relay their experiences. He not only found details about other people's treatments, but lots of support and comfort. He was able to make a decision based upon his own research and knowledge.

The sheer volume of health information on the internet changes the dynamic of health information dissemination. In 2021 there were over a billion websites on the internet and many millions of blogs. Of those, several million websites, more than one million blogs dealt with health-related information. These sites include health content that provides you with information for staying well, for preventing and managing disease, and for making decisions about health, healthcare, health products, or health services, as well as information that can help you avoid a major medical error or misdiagnosis.

At age fifty-two, Trisha found a large lump on her torso. She immediately saw her primary care physician, who sent her to a surgeon who removed the lump and sent it out for diagnosis. More than two anxious weeks went by before Trisha heard back from the surgeon, who finally told her that she had a very rare form of cancer for which there was no cure. The surgeon explained that with chemotherapy, she might gain two years of life. Trisha immediately accessed the internet for information while she waited to see an oncologist, who was not much more encouraging than the surgeon.

However, the oncologist did send Trisha for blood work and a CT scan. Both came back negative for any abnormalities. Trisha was feeling fine; with the mixed the test results, she suspected pieces were missing. While waiting to see the next doctor, she continued to search the internet for help interpreting her tests. By the time she saw the second oncologist, she was empowered with all the necessary information. This doctor agreed that there were significant doubts about her diagnosis and, with her permission, sent the test results to the National Institutes of Health. Three weeks later, she received a notice that, indeed, she did not have the cancer.

Trisha went through an agonizing period. Because she is an e-patient who persisted in searching for every possible clue and learning about the experiences and opinions of others, Trisha was able to avert an alarming misdiagnosis that could have led to unpleasant and risky chemotherapy and even death. Although Trisha's web searches did not provide absolute answers, the information she found set up enough red flags to make her suspicious and able to convince the second oncologist to pursue the issue. This approach turned out to be exactly correct.

Information Overload

Information access is a good thing because it ensures that you will be a more informed, healthcare-literate patient who can participate completely in taking care of yourself and your family. However, the sheer amount of health-related information present on the internet and in the media puts you in extreme overload. As a result, it is difficult to filter what is credible and reliable and what is hype. Below are guidelines regarding what to look for on websites that you can rely on to provide you with the healthcare information you need.

1. **Questions to Ask**

   - Who developed this site? Are the author or authors clearly identified? Are the credentials of the author listed?

- Does the page show when it was last updated? Are the links to other resources still active?
- Is contact information provided so that you can email, call, or write the web sponsor?
- What is the purpose of the information?
- Can the information be verified in other sources?

2. What are the trustworthy websites?

During the end of the twentieth century and early part of the twenty-first century there were organizations that gave a stamp of approval to credible web sites. Today with the 1. 88 billion websites and increasing daily by the hundreds, ensuring the credibility and reliability of information on websites is not really possible. However, there are some websites that you can trust because they are generated by established, known organizations, including:

- Websites of established healthcare organizations, government agencies, (e.g., HHS, CDC, and NIH), which can be held to the test for accuracy, reliability, and impartiality.
- Sites of the major medical institutions including: Johns Hopkins Medical Center, Partners, Kaiser Permanente, and Mayo Clinic. During the Pandemic, Johns Hopkins Medical School affiliated researchers posted statistics daily in the press about where the pandemic was most prevalent and how many cases and deaths occurred throughout the country.
- Websites of organizations that are focused on specific diseases, such as the American Cancer Society, American Lung Association, and the Brain Tumor Society, are excellent resources for information on these conditions and also a great place to find individuals who are experiencing similar problems with whom you can interact. These sites have a vested interested in helping you and your physicians find information of value.

When you are interested in purchasing health-related products, you should choose reputable websites that require that you have a prescription written by your health provider. They strictly obey the regulations enforced by US government agencies. Many sites on the web that sell you drugs were developed outside the United States where there is no federal agency such as the FDA monitoring the research and distribution of drugs.

A study reported in the *Journal of the American Medical Association* reviewed 443 websites that contained information on eight best-selling herbal products. Researchers found that many sites were selling the supplements, and over half of these contained illegal and misleading claims that their products can treat, prevent, diagnose, or cure specific diseases. Many of the websites analyzed contained illegal health claims and information that could potentially harm. What was more alarming about the findings is that more than half of the respondent internet users reported that they thought "almost all" of the health information they encountered on the internet was credible.

Good websites state their purpose on their home page and include search options for easy navigation. Always check to see that the information is current. You want to find sites that focus on content that meets your specific needs. Many sites include chat rooms and message boards where you can post your questions and share personal information that can be valuable.

General Purpose Health Websites

www.medlineplus.gov

This portal site, sponsored by U.S. Federal Government branches including the National Library of Medicine and the National Institutes of Health, is based upon reliable, medically approved information. Over three thousand doctors, nurse practitioners, and editors contribute content to Medline plus, which is available in both English and Spanish versions. It includes photo illustrations an extensive drug database and over four thousand articles about disease, tests, and symptoms, as well as a medical encyclopedia and a database of doctors, dentists, and hospitals, and clinics. The Medline plus online directory of drugs also includes a database on herbal supplements. The entire site is a great resource for the healthcare professional; however, portions of this site could be technical.

www.cdc.gov

This website was developed by the Center for Disease Control (CDC) to provide all of the stakeholders in healthcare with information on public health issues. One of the mandates for the CDC is to maintain communications with state and local government agencies, federal government partners, and the general public. The website is the major communication channel to accomplish this and includes travel alerts, vaccine requirements, food and beverage, swimming and insect warnings, and disease outbreaks. The CDC website also sponsors podcasts and publications. Discrete sections of information on the site are specifically targeted to consumers. The CDC's specific formats for communication enables it to provide the appropriate assistance and information to the general public.

www.healthfinder.gov

Healthfinder links to a broad range of consumer health and human service information resources. Healthfinder was developed and is maintained by the US Department of Health and Human Services. It allows you to submit web searches for medical, pharmaceutical, or healthcare information and receive targeted useful health information free of charge. This gateway website links to selected online publications, clearinghouses, databases, and over 1,500 websites. It provides information on every conceivable health topic.

www.webmd.com

WebMD is probably the most popular health website and includes timely and relevant news, information, educational materials, video clips, message boards, and healthcare blogs, all of which are there to engage the e-patient. At WebMD, you can create a personal health record and search for a doctor or hospital, or use several worthwhile tools for information on weight and diet, a BMI calculator, advice on dental issues, skin and beauty tips, videos, a symptom checker, and more.

www.google.com/Top/Health/

The Google search engine is based on the Open Directory Project (DMOZ, directory.mozilla.org) and is a multilingual, open-content directory of web links that are maintained by a community of volunteer editors who keep the content up to date and relevant. The search engine includes Harvard Medical School's consumer health information and journal databases and a complete spectrum of health topics for individuals of all ages. Subjects covered include: men and women's health, conditions and diseases, health research, informatics, and general consumer health concerns such as weight loss, travel health, nutrition, insurance issues, fitness, cosmetic surgery, medications, and many more.

www.familydoctor.org

This site from the American Academy of Family Physicians features comprehensive consumer health information in both English and Spanish. Some of its interesting features include the "Smart Patient Guide" to help you advocate for your own healthcare; a guide to over-the-counter medications; a list of parameters to help you choose a family doctor; and Health Tools, a section that takes you to links on conditions from A to Z, with a search engine by symptom and disease. At familydoctor.org, you can also find answers to insurance questions, Medicare Part D, and other public policy issues. There is a link titled "Advocacy" that enables you to send messages to Congress on healthcare issues you are concerned about.

Finding Doctors and Hospitals

There are several websites where you can search for information about doctors and hospitals; some are more credible than others. Most of the gateway sites listed above also include search parameters for finding doctors and hospitals by geography and by subspecialty. Listed below are specific sites that are reliable, credible, and frequently updated.

Doctorfinder (ama-assn.org)

This comprehensive web database directory of medical doctors and osteopaths practicing in the United States is sponsored by the American Medical Association. At this site, you can search by physician name or specialty. Information provided includes basic professional data such as medical school, year of graduation, board certification, telephone number, and address.

If you are looking for a hospital by a particular location, this website provides you with a search engine that will get you the name of the hospital in that particular geography. From that information, you can go directly to the hospital's website and evaluate whether or not it meets your criteria for size, specialized care area, etc. It is a good way to begin your quest.

Institutional Websites

One of the first places that you will turn for health information is the website of the specific hospital or medical institution that you and your doctors are affiliated with. These sites offer not only information to help you navigate the system but general information on health issues as well. Among the more comprehensive institutional sites are those sponsored by health insurers (e.g., Humana, Blue Cross, Aetna, and Kaiser Permanente).

Professional Organizations

For patients who have a specific health problem, the best sources of information and interactive discussion groups are the sites sponsored by the associations that support those diseases

Online Resources for Cancer Patients

www.cancer.gov

The National Cancer Institute (NCI) was among the first federal agencies to recognize the potential of the internet for disseminating health-related information. The NCI site is user driven and offers cancer patients stakeholder meetings, focus groups, standard and customized online user surveys, and usability testing. The site includes information on every type of cancer and discusses new treatments and clinical trials of new drugs. As cancer patients increasingly turn to the web in large numbers seeking to be informed decision makers in their own care, NCI has responded by organizing the Cancer Net website by audience type with entry points for patients, health professionals, and researchers.

www.cancer.org

Cancer patients use internet resources for general information about the disease as well as for pointers on how to secure a second opinion and how to interpret issues presented by health providers. For patients with rare forms of cancer, the internet often is the only way to get support and practical advice. This site includes the hard-to-find information on the financial impact of cancer and work-related consequences. It also lists organizations offering financial assistance and provides guidelines for those of you who need to become familiar with insurance coverage, how to submit insurance claims, and help with keeping records.

www.cancercare.org

Cancercare is a nonprofit organization that provides free professional support services for individuals afflicted with cancer. At this website, there are a number of online support groups as well as education workshops where you connect in by telephone and listen to discussions on various aspects of cancer. There are also podcasts related to medical, emotional, and practical concerns of cancer patients and a comprehensive online handbook that provides financial assistance resources sorted by the type of cancer and the area where you live, within a ten-mile radius of your zip code.

Online Resources for Diabetes

Diabetes is a disease in which blood glucose levels are above normal. It is one of the fastest growing health challenges of the twenty-first century. Over 37.3 million Americans—about 1 in 10—have diabetes. About 1 in 5 people with diabetes don't know they have it. Individuals who are diagnosed with diabetes have more than tripled over the past twenty years. Studies have shown that online programs do work. Additionally, there are hundreds of diabetes websites that offer suggested menus, programs, and products that are not medically approved and need to be approached with caution. There are two credible information resources on the web where you can learn everything you need to know about diabetes beyond what you learn from your own doctor.

www.diabetes.org

This site is the home site of the American Diabetes Association. It includes information for individuals with both Type 1 and Type 2 diabetes. It provides useful and reliable information in how to live healthy with diabetes and lists of tools and support programs, an extensive guide on products for the diabetic and information on medical reimbursement programs.

www.cdc.gov/diabetes

This site, which is a section of the CDC site described above, reviews symptoms, types of diabetes, risk factors, treatment options, prevention, and advice.

Web Resources for Cardiac and Lung Diseases

www.americanheart.org

This website is divided into sections for patients, caregivers, and professionals, and is full of information on how to deal with cardiac disease as well as tips on how to stay heart healthy. The Health Hub section is a patient portal where you can find tools and resources about cardiovascular disease and stroke, including a risk assessment, as well as comprehensive information to help you understand and manage your health and to determine your treatment options. Information resources include articles, podcasts, and videos. The website also has a complete description of every potential heart health condition, from arrhythmias to cholesterol.

National Heart, Lung, Blood Institute (NHLBI)

At this website from the National Heart, Lung, and Blood Institute, you can find more information than you ever thought possible on heart attacks, high blood pressure, obesity, cholesterol, asthma, emphysema, lung cancer, and other issues related to heart health and lung illness. Health assessment tools are also available on this site so that you can take a more interactive role in learning about specific issues that might affect you. If you join the NHLBI Health Information Network, you will receive periodic emails about the latest key findings, research information, articles, and educational programs.

Health Websites for travelers

If you are planning a trip outside the United States you will want to know what to expect about air, food, water or bacteria you might encounter that could adversely affect your health. In some cases, you need vaccines when traveling to certain parts of the world. The CDC Travelers' Health site, www.cdc.gov/travel should be your first stop. Here you can find information about:

- Vaccine requirements
- Food and beverage precautions
- Swimming precautions
- Insect precautions
- Disease outbreaks (CDC Blue Sheet)

You should also check the current State Department warnings before travel that you can access at https://www.state.gov/travelers/ where you will find extensive emergency information and resources to help guide you through issues that may occur prior to leaving the country and particularly while you are away.

Before you leave for a long trip, you might want to locate a doctor or appropriate medical facilities in a specific geography, particularly if you have any type of chronic condition. Two organizations can offer your assistance:

The International Association for Medical Assistance to Travelers www.iamat.org, and the International Society of Travel www.istm.org

Online Pharmacy Sites

According to Pew, nearly one-half of American adults, or about ninety-one million people, take prescription drugs on a regular basis. One in four Americans have looked online for drug information. Nevertheless, the same Pew study found that only a fraction of Americans has ever bought a prescription drug online. That is slowly changing as the cost of drugs continues to rise and access to new drugs is limited and delayed by health insurers and the US Food and Drug Administration (FDA) clinical trial and approval process. Note that all reputable sites require that prescriptions be written by a patient's health provider and adhere to regulations enforced by government agencies such as state pharmaceutical boards and the FDA. However, prescription drugs can and are being purchased at sites located in other countries where there are few if any laws or enforced policies. If you are tempted by lower drug prices and easier access on a website that may be from another country, recognize that those drugs not approved by a US regulatory agency can be compromised in dosage as well as formula and may contain contaminants, other ingredients, and even other types of drugs. You want to take proper care to ensure that the drugs you are taking are in no way injurious to your health. Although no drug is 100% safe, best practice teaches us to order only from online pharmacies that adhere to the industry's self-regulation mechanisms, such as the Verified Internet Pharmacy Practice Site, VIPPS, run by the National Association of Boards of Pharmacy.

The FDA, www.fda.gov publishes a warning for consumers buying medical products online. The site includes a list of all drugs. If you click an individual drug, you will go to a page that will give you a complete overview of the drug and its adverse effects. For example, if you click on the over-the-counter drug naproxen, you will go to a section of the website that links you directly to a patient information sheet, a healthcare provider information sheet, a prescription drug naproxen label, a paper that was issued as a public health advisory recommending limited use of Cox-2 inhibitors, and a public health advisory on nonsteroidal anti-inflammatory drug products (NSAIDS). Naproxen is one of many drugs on the market in this category.

Social Networking and Health

In the US today, eight in ten internet users search for health information online, and 74% of these people use social media. E-Patients are using social networks to ask about health and find others with the same condition at the same stage of treatment to talk about common illnesses, as well as more complex conditions. Social media is effective as a way to spread awareness about what a specific organization can offer patients in terms of medical care. It can inform the public about what advanced technology or treatments a facility offers. It is also one of the most cost-effective distribution tools for healthcare information, however it is essential that patients understand that the information is not vetted and can sometimes be erroneous. Issues you want to be aware of regarding social networks include:

- If you are going to use social networking to resolve health issues or learn more about health problems, determine your goals.
- Stay focused on your own personal needs and do not let fascinating, friendly people deter you from your main interest.
- Spend some time seeking the specific sites that are most useful to you. Once you find a group that has relevant information for you, be sure to participate in the conversation and share information

www. Facebook.com

Facebook's ability to connect you to a vast community of "friends" makes it easy to pose questions and find common ground on health issues that may be bothering you. You can easily create your own subgroup or join groups that have formed around health issues.

www.twitter.com

On Twitter you can engage in quick exchange of information using one simple statement of 150 characters or less. You transmit that message to fifty, five hundred, or five thousand of your colleagues, friends, or associates, provided they have a Twitter account and are following you. Your message can be a question about a health issue, a comment on a health experience, or an appeal for assistance. You can also use Twitter to follow experts or individuals who may have information on similar health problems and who tweet about their experiences or offer advice. Although this may not be clinically useful, it may help you become educated. Both Twitter and Facebook help you brainstorm health problems that may be on your mind.

www.patientslikeme.com

PatientsLikeMe.com allows an e-patient to connect with individuals by disease or community and share data on treatments, medications, and outcomes. Active communities include ALS, multiple sclerosis, Parkinson's disease, HIIV/AIDS, many areas of mental health, and many rare diseases that do not have other websites devoted to them. New disease communities are being added constantly, and at PatientsLikeMe.com, you are invited to start a community if you have some special disease you want to talk about.

www.Inspire.com

This is a leading social network for patients with health issues that includes over 200,000 users and 200,000 diseases that they cover, with special discussion groups, moderated by trained patient advocacy specialists. There are 8,000 posts on inspire covering a wide breadth of health concerns Members connect across time and space: around the globe 24 hours a day/ 7 days a week, sharing health information and support without regard to location, background or status. Together they reduce the feelings of isolation experienced by those with chronic conditions, cancer, and rare diseases.

www.grouploop.org

Group Loop is a social network for teens diagnosed with cancer that provides a place where teens can find individuals to talk with about their condition and treatment and open up about their emotional and social concerns while they are going through this scary, difficult time in their lives. They can also find lots of educational tips about cancer. Since so many teens have already connected through social networks such as Facebook, Group Loop has had very wide appeal. Content includes "In the Loop," a personal profile posted each month and relayed by that individual. Another section on the website offers parents advice on dealing with their own stress and helping their teens get through this time.

www.myfamilyhealth.com

At MyFamilyHealth.com, you can create a family tree, record your family health history, and assess your health risks by doing the following:

- Creating a medically useful record for the entire family.
- Discovering and learning about health problems that run in your family, including conditions that might not otherwise be considered by your physician.
- Discovering if you or your family members could benefit from specific diagnostic, genetic, or screening tests.

Tim Berners-Lee, Director of the Worldwide Web Consortium and one of the founders of the web, described the internet as "a universal space of information, designed to enable human communication through shared knowledge." Having the internet available enables you to be more informed, more in touch, and more involved with decisions that affect your life. Current technology has added a new element—collaborative virtual communication. As an e-patient, you have a responsibility to use this information with good judgment, keeping confidential the information that you and others share through online communities and social networking and maintaining the privacy of your own health information.

Sites to Avoid include:

Websites that offer suspicious remedies that promise fast weight loss, instant hair replacement, miracle cures, nutritional supplements, great healing.

- Websites that advertise doctors who will supply a quick fix.
- Websites that are filled with medical jargon.
- Websites that try to sell you products, vitamins, health services, common drugs, or herbal remedies.
- Websites that are overly negative.

Advice, Diagnosis, and Drugs

There are many sites on the internet that now offer medical consults and diagnosis that are typically not covered by insurance. It is important to be vigilant about who is giving the advice, how well that individual knows you and your individual problems and health issues, and how valid the instructions are when compared to what your primary care provider might tell you. It is risky to consult a doctor who has not seen you in person at least once during the past twelve months or who does not have access to your health history, medication list, electronic health record, and other facts that are part of who you are.

Key Points

1. As an e-Patient, you are a seeker of the good information. Use the sites listed in this chapter as a guide because they are reliable, credible, and frequently updated to provide you with the information you need.

2. Your physician should be your primary resource for direction on how to use the internet for medical research and what web sites you should check out that have useful information on your particular health profile. Ask your physicians to supply you with a list and set up a plan to discuss questions after your web research is completed.

3. Be careful about the health products you are tempted to purchase on the internet. Check out the contents of products to find out whether they are FDA approved and offered by a credible source.

4. Do not overlook the websites sponsored by your health insurance provider. Do not hesitate to use the customer service department for clarification regarding medications, treatment plans, questions, or observations.

5. Social networking sites offer you the opportunity for collaboration and interaction, which can help you with healthcare questions and concerns. On the other hand, these sites now advertise medical products as well as other consumer goods and use tracking tools to identify what you are looking at and purchasing. Be aware that this type of monitoring could invade your personal privacy. Use these sites with caution.

6. The internet is global, and there are many healthcare sites that will try to sell you products and services that are not FDA approved and may not be in your best interest. Be sure that you are exercising good judgment. When in doubt, consult resources that you can trust.

Table 3
Best Websites

Purpose	URL	Description
Search Engines	www.google.com/ tophealth/	Multilingual open content directory of web links with constantly updated content
Government Portal Site	www.medlineplus.gov	Input from over 3,000 medical professionals; includes disease and drug database; photos; in English and Spanish
Public Health for consumers	www.cdc.gov	Travel alerts and information; public health announcements, educational materials, podcasts
General Consumer information (government sponsored)	www.healthfinder.gov	Gateway to consumer health information from variety of government sources
General Consumer information (privately sponsored)	www.webmd.com	Timely news, information, educational tools, option to create PHR, podcasts, message boards
Finding a doctor or hospital	doctorfinder.ama.assn.org /doctorfinder	Database directory of doctors practicing in US including basic professional information
Institutional website	www.mayoclinic.org	Information about illnesses, drug listings, discussion groups, tolls to monitor disease, chat with specialist, links to health news
Cancer	www.cancercare.org	Free professional support services; online support groups; workshops
Cardiac Issues	www.americanheart.org	Tools and resources about cardiovascular disease and stroke including a risk assessment and tools to manage heart disease
Social network	www.patientslikeme.com	An open forum where patients share health experiences, learn about the latest treatment options; covers most disease types

Notes

1. Susannah Fox and Sydney Jones "The Social Life of Health Information," Pew Internet and American Life Project and the California Health Foundation, June 11, 2009, pp. 2–4.
2. Susannah Fox, "Health Topics," Pew Internet and American Life Project and the California Health Foundation, February 1, 2011, p. 2.
3. Ken Euteneier, Founder of the website Fibrohugs, in an interview with the author. For more information go to http://fibrohugs.com
4. Newsweek Magazine, April 25, 2005, p. 62.

CHAPTER NINE
Protecting Your Privacy

When dealing with data, scientists have often struggled to account for the risks and harms using it might inflict. One primary concern has been privacy - the disclosure of sensitive data about individuals, either directly to the public or indirectly from anonymized data sets through computational processes of re-identification.

Kate Crawford, Australian, born 1976, writer, composer, produced, academic, thought leader.

Confidentiality and Information Flow

There was a time when your personal medical information involved a conversation between you and your chosen medical provider, and no other person was involved. In modern medical practice, where numerous tests, consultations, and procedures are routine, and referral to specialists by a primary care doctor is standard practice, your information is accessed by numerous faceless institutions and strangers involved directly or indirectly in the diagnosis and treatment of your medical conditions or in the payment for your care. Each time you visit a doctor, therapist, hospital, or clinic, information about you is recorded and filed either on paper or, more commonly today, on a computer. Typically, this information includes a description of your symptoms, your medical history, examinations, medications, test results, diagnoses, treatment, payment data, benefits, health plans, referral for the visit or procedure, and the authorization for it to happen. Your social security number, birth date, and other relevant personal information may also be included. Third-party reimbursement, use of email, e-prescribing, health information exchange, electronic medical and personal health records, and other e-health initiatives raise questions about how easily that health information can be penetrated and tampered with by unauthorized individuals.

A patient at the Brigham and Women's Hospital in Boston learned that while she was an inpatient, various medical staff at the hospital had accessed her medical record more than two hundred times. Although this occurred before the passage of HIPAA and the HITECH Acts, that restrict access to your personal health information without your express, written consent, there are many instances when your health information is viewed by many eyes, and you may not be aware of just who is doing the viewing. For example, when you go to the emergency room of a hospital, a record is created that goes with you wherever you are transported. While the purpose is to ensure continuity of care, this record is handled by many healthcare workers, including doctors-in-training, attending physicians, nurses, and technicians in all of the labs that you visit. On the general medical ward in the hospital, your record is seen by nurses, nursing assistants, residents, attending physicians, and specialists who are involved in your care. At your health plan, your health information is exposed, processed, and recorded by numerous individuals. So how do you ensure that your information is accessible only to those authorized to have access? There is no easy solution. (1)

Dave saw his primary care physician because he was having problems with his lower back, and it seemed to worsen each time he drove his car. The doctor ordered an MRI and suggested that Dave see an orthopedist. When Dave visited the specialist, he learned that two of the disks in his back were bulging and one was herniated. The doctor recommended a series of steroid injections and sent Dave to a physical therapist. Each of Dave's doctor's visits, PT sessions, and tests were billed to Dave's health plan. Three months later, Dave received a letter on health plan letterhead providing helpful suggestions on how to manage his lower back problem. Without Dave's knowledge or consent, his health plan had forwarded information about his back problem to a disease management company intended to help him. Dave was so impressed with this personalized information; he did not stop to think about whether or not he had signed permission for anyone outside the health plan to view his medical information.

Although legally this type of sharing is permitted, as long as the disease management company is acting as a business associate of the health plan, had Dave's condition involved a more sensitive issue such as mental health, HIV, drugs, or alcoholism, he probably would have been upset by a third party being privy to his personal health information.

These examples raise questions about the confidentiality of your health information. every e-Patient should consider the following:

- How is my health information being used?
- Should I worry about the availability of my health information to so many providers and healthcare workers at institutions where I receive care, as well as to individuals at my health plan, pharmacy, and other organizations that cover my care?
- Do electronic health records result in more unauthorized eyes viewing my information?
- How do I monitor what happens to my health information that is stored online (perhaps in a personal health record or in my providers' databases)?
- What can I do about maintaining the privacy of my health information?
- What recourse do I have if my privacy is violated?

Privacy is defined as "excluded or isolated from view or contact with others." Confidentiality, which should be the basis of your relationship with your doctor, is defined as "ensuring that your health information is accessible only to those authorized to have access." Often there are overriding legitimate reasons why health information must be shared, for instance, to avert medical error, to ensure your safety, to provide you with the necessary treatment, and to ensure that society is properly protected. There are cases where the courts have held that doctors have an obligation to pass on information to protect patients, for example, when the doctor must warn police of a mentally unstable patient who has threatened a specific person with violence. The current health reimbursement system requires that doctors transmit medical information to payers as a condition of coverage and as a safeguard against fraud and overpayment. In most instances, your health information is sacrosanct and your privacy is maintained; however, that is not always the case. What are your protections?

HIPAA

There are laws regarding the privacy of your health information. In 1996 the United States Congress enacted the Health Insurance Portability and Accountability Act (HIPAA). Most consumers, and many healthcare providers, do not understand the fine points of HIPAA.

Title I of HIPAA protects health insurance coverage for workers and their families when they change or lose their job.

Title II of HIPAA, the regulations, titled, Preventing Healthcare Fraud and Abuse, Administrative Simplification Medical Liability Reform, defines numerous offenses relating to healthcare and outlines civil and criminal penalties. It creates several programs to control fraud and abuse within the healthcare system. The most significant provisions of Title II are the administrative simplification rules, which establish national standards for the use and dissemination of healthcare information stored on paper or in a digital format. These rules also apply to electronic healthcare transactions and national identifiers for providers, health insurance plans, and employers. They address the security and privacy of health data, particularly health data that exists in digital format. Since the number of consumers who connect with physicians online has increased since the beginning of the twenty-first century, HIPAA has gained greater importance.

The Privacy Rule

The HIPAA Privacy Rule regulates the use and disclosure of certain information held by "covered entities," generally healthcare clearinghouses and employer-sponsored health plans, health insurers, and medical service providers that engage in certain transactions. It went into effect April 2003 and established regulations for the use and disclosure of Protected Health Information (PHI). PHI is any information concerned with health status, provision of healthcare, or payment for healthcare that can be linked to you. PHI consists of any part of your medical record or payment history; It also includes a provision that outlines your right to access your medical records and control how your personal health information is used and disclosed. A key premise of the Privacy Rule is that your health information must be properly protected, while allowing the flow of health information needed to provide and promote high-quality healthcare to the public. Below are specifics of the Privacy Rule that are important for you to understand: (2)

- **Privacy practices notice:** Your providers, with certain exceptions, must ask you to read and sign a notice of privacy. This notice describes ways in which your healthcare professionals may use and disclose your protected health information and includes an explanation of your right to complain to your provider and/or the Department of Health and Human Services (HHS) if you believe your privacy rights have been violated. (2)

- **Patient access to medical records:** Upon request, you must be given control over how your personal health information is used and disclosed. You have the right to see your medical information and to copy and supplement—but not to change—your medical record. You also have the authority to refuse to have certain individuals or organizations see your records.

- **Appropriate safeguards:** The Privacy Rule states that healthcare providers, health plans, and information clearing houses that collect, share, and store health information must have appropriate technical and administrative safeguards in place to protect that information. Additionally, alternative means of communicating health information such as email may be used only with appropriate safeguards.

- **Limits on use of personal medical information:** The Privacy Rule sets limits on how health plans and providers may use your health information. There are no restrictions on your doctors, nurses, and other providers in sharing information needed to treat you. However, personal health information may not be disclosed to your employer without your written consent, nor may it be used for purposes not related to healthcare. Your information may be shared with a family member or other designated close friend or relative with your consent. That individual may act on your behalf to pick up prescriptions or medical supplies and provide other needed health assistance.

- **Limits on use of health information for marketing purposes:** The Privacy Rule sets restrictions requiring you to sign a specific authorization before your providers or payers can release your medical information to a life insurer, a bank, a marketing firm, or anyone else not directly involved in your care. While doctors and other providers and payees are permitted to communicate freely with you about treatment options and disease-management programs, marketing of your health information is not allowed.

- **Confidential communication:** Under the Privacy Rule, you can request that your doctors, health plans, and other providers take reasonable steps to ensure that their communications with you are kept confidential. For example, you can ask a doctor to call you at a specific location rather than at home, and the doctor's office will comply with that request if it can be reasonably accommodated.

- **Complaints:** If you have concerns about privacy, you may file a complaint using the HIPAA Health Information Privacy Complaint Form developed by the Health Privacy Project. These forms are available at HealthPrivacyProject www. healthprivacy.org. You may also email a complaint to the Department of Health and Human Services in Washington, D.C, at, OCRcomplaint@hhs.gov information on filing and updates to the Privacy Rule can be found at www.ocr/discrimhowtofile.

The enactment of the American Recovery and Reinvestment Act of 2009 (ARRA), effective November 30, 2009, extended the complete Privacy and Security Provisions of HIPAA to business associates of covered entities, including the extension of newly updated civil and criminal penalties to business associates. Other provisions of ARRA (Section 13402–13406) tightened the privacy rule. If a business that has possession of your health information experiences a breach involving the unauthorized acquisition, access, use, or disclosure of your unsecured, protected health information that results in financial, reputational, or other potential harm to you, it must notify you or call or contact governmental authorities. This includes online companies that house your personal health record, such as WebMD, Microsoft Health Vault, and Google Health, if they are business associates. Additionally, the rules regarding the marketing and communicating of products or services to you based on your health information were tightened.

The changes brought about with the passage of ARRA also require your providers and their business associates to honor a request by you to restrict disclosure of protected health information to a health plan if you have paid in full for a specific service (e.g., a second opinion) and the health plan is not involved in the reimbursement of that particular service. Further, Section 13405(d) of ARRA prohibits any direct or indirect remuneration in exchange for your protected health information without your authorization and further restricts the marketing of your health information or the marketing to you of products by vendors based on information they have obtained from your health records without your explicit permission. There are also provisions in Section

13408 of the ARRA stating that you must provide permission for transmission of your records among entities in regional health organizations, health information exchanges, or e-prescribing networks. (3)

While these federal laws and rules provide a broad umbrella under which the privacy and confidentiality of your health information is protected, state laws and common law (case law) offer even more powerful safeguards. Although there is great variance from state to state, these laws are generally more specific regarding the following:

- Written consent for the sharing of medical records, especially information that relates to sensitive medical conditions and health issues
- Issues that deal with the rights of parents and minors
- Stronger protection for specific diseases and conditions, especially HIV, mental illness, drug addiction, and alcohol addiction
- State data breach laws if your social security number of other confidential information is disclosed without authorization

Online Storage of Health Information in

In twenty-first century, healthcare, the increasing use of the internet for the collection, storage, and communication of medical and other personal health records enables agencies, institutions, and strangers throughout the world to view your medical records or medical information. Large databases that include these records are stored on servers connected to and accessible via the internet. Some physicians and patients use the internet without a secure portal to send and receive email messages containing health information. Compounding the problem is the fact that many patients are required to provide personal health information to gain employment, procure insurance, obtain a credit card, or participate in today's economy. The bureaucratic institutions that control and use that information offer virtually no accountability or rationale that explains why or how the information will be used or stored.

Electronic health records are generally accessed using a password protected system. Individuals with a need to see the record are typically granted access to only selective parts of the record. Audit trails closely monitor who has viewed the record. However, even with these protections, the same electronic information that sits behind secure firewalls and is encrypted has already been subject to tampering. Additionally, there is the issue of the privacy of email communications that you elect to send, even though they include private health information. There have been many instances where patient information is shared with every good intention, by providers trying to manage chronic disease issues of patients. The benefit that accrues from these activities to the patient population can outweigh any threats to privacy, but it is another slippery slope that must be approached with care. In many clinical studies, patients will provide permission for their data to be used. Generally, there is a real effort to scrub that data so that names and specific identification are not possible. However, there is always the chance that information could be traced back to an individual.

The Centers for Medicare and Medicaid Services (CMS) has amassed databases that gather information to aid health plans, physicians, and providers in an effort to improve quality and to help Medicare beneficiaries and other consumers and health plans make informed choices. These "quality initiatives" are based primarily on information gathered directly from patients. The data is scrubbed so it is not likely that anything could be traced to specific information about individuals, but privacy issues loom whenever government collects that type of information, even if it is intended to benefit the public. (4)

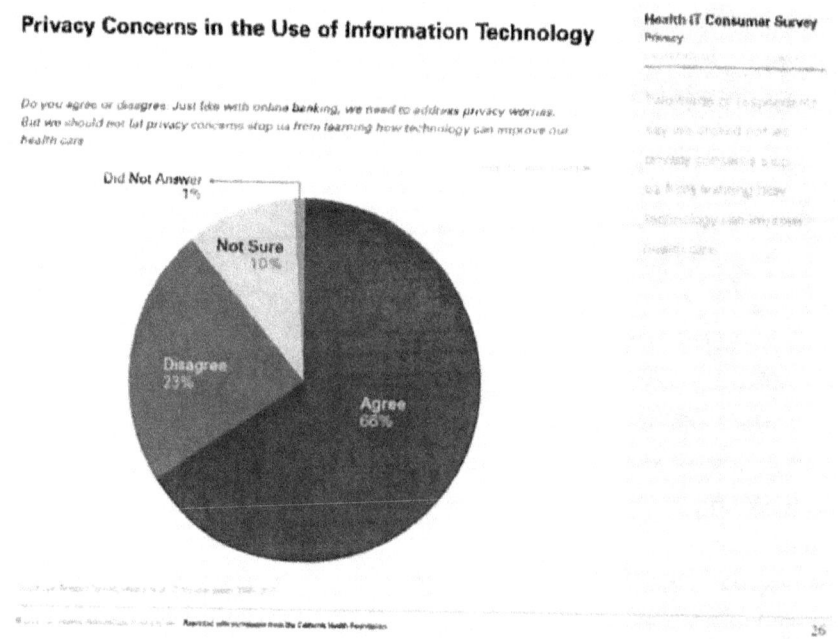

Privacy Concerns in the Use of Information Technology

The proliferation of USB devices such as thumb drives, flash drives, and MP3 players creates a situation where data can be downloaded to these tiny devices and stolen with the click of a mouse. Patient files, folders, and personal information residing on the computers at HMOs, hospitals, and other health-care organizations are at risk for invasion of privacy Even the federal government, the entity that physicians and patients assume to be the overseer and protector of the public's health information, has violated that trust. There have been several instances where attorneys for the United States Justice Department attempted to thwart privacy protections and obtain medical records. These actions have been directed at the records held by Planned Parenthood clinics related to women who had or were planning abortions. The records sought were paper, and the Justice Department argued that the names could have been removed to protect the identification of the individuals involved. But the question remains whether or not an electronic medical record makes it easier for the government to pry. (6)

The CVS Case

CVS implemented a Patient Compliance Program to provide healthcare information to its pharmacy customers. Using a database prepared from customers' prescription records, a third-party company mailed letters on CVS's behalf to several of its pharmacy customers. Those customers were not previously informed of or asked to consent to this program. The mailings concerned drugs and medical conditions and took a variety of forms. Some letters provided information on the risks of certain health conditions; some letters urged customer to switch to a new or alternative prescription medication; one mailing addressed the benefits of taking and refilling prescriptions. Each letter included a box to check if the customer did not want to receive further mailings. Although the letters were written on CVS pharmacy letterhead, the mailings were financed by several drug manufacturers.

Jeffrey Kelley, who use CVS for obtaining his diabetes medicine received a letter that talked about the dangers of high cholesterol, a known risk for diabetes patients. The letter was clearly marked as being sponsored by Merck and Company, Inc., a manufacturer of the anti-cholesterol prescription drug Zocor. Kelley felt that the letter invaded his privacy and, along with other individuals who had received similar letters, initiated a class action lawsuit against CVS in Massachusetts, and the pharmaceutical companies who had participated in the program. The lawyers argued that CVS and the pharmaceutical companies, for their own financial gain, improperly used confidential medical information provided to CVS by its pharmacy customers, thus violating the privacy rights of those customers.

The court held that the letter did not violate Mr. Kelley's right to privacy. In Massachusetts, an invasion of privacy requires an "unreasonable, substantial, or serious interference" with privacy. In this case, CVS did not disclose Mr. Kelley's name to Merck. They merely provided his date of birth, and address to the mail house. They did not know why Mr. Kelley had been selected to receive the letter and did not receive any information about his medical or pharmaceutical history. The court reasoned that there is nothing improper about a pharmacy reviewing its prescription database and providing relevant information to its customers. The fact that CVS used a third party to do this was essentially the same as hiring extra clerks to send the information directly from the pharmacy. The court also noted that Mr. Kelley "readily disclosed" his diabetes to friends and associates, which made it "even clearer that the information disclosed by CVS cannot reasonably be deemed a substantial or serious interference with his privacy." The court, however, did find that CVS engaged in an unfair and deceptive act by concealing the fact that it profited from sending the letters. (6)

Privacy Dilemma in the Technology Era

As a society, we cannot afford to halt progress by refusing to use electronic technology or disallowing data to be electronically stored. However, as we see from these examples, the risk that the data could be abused never quite goes away. In 1977, a group of patients and prescribing doctors brought an action challenging the constitutionality of a New York State statute that gives the state the right to maintain a database of patient names, addresses, and ages of people who use the most dangerous legitimate drugs. This data is filed with the New York State Health Department and kept for a five-year period using a system that is supposed to safeguard the security of the information, after which it is destroyed. Public disclosure of the patient's identity is prohibited, and access to the files is confined to a limited number of health department and investigatory personnel.

A case regarding the existence of this database, and the privacy questions involved, was heard by the United States Supreme Court (Whalen v. Roe, 429 US 589, 1977). In a unanimous decision, the court decreed that the New York program does not, on its face, pose a sufficient threat to establish a constitutional violation. The court noted that that the state's carefully designed program includes numerous words and safeguards intended to forestall the danger of indiscriminate disclosure.

> We hold that neither the immediate nor the threatened impact of the patient identification requirements in the New York State Controlled Substances Act of 1972 on either the reputation or the independence of patients for whom Schedule II drugs are medically indicated is sufficient to constitute an invasion of any right or liberty protected by the Fourteenth Amendment. (7)

In a perfect world, we would have the efficiency and economies that digital technology brings to information management, collection, and storage, as well as the privacy protections that always ensure that the data cannot be penetrated or disturbed. However, it is not such a perfect world. The digital accumulation of data has moved at a much faster pace than the ability of the same digital technology to protect that data. As these cases show, there are also many gray areas where privacy issues emerge. Therefore, the best you, an empowered patient can do, is to use all the available protections and surveillance systems, while living with the constant awareness that there are risks of your health information being disclosed or used by persons without your knowledge or consent.

What can you do to help protect your privacy? Consider the following:

- Have you provided clear instructions to your providers regarding how you want your information to be used and disseminated regarding other medical professionals they might confer with, your family and friends, your health plan, and pharmacists?
- Have you taken the proper care and considered the privacy of your health information when you put data online, particularly in respect to social networks?
- Do you assiduously avoid providing your social security number in your medical records and insist on giving out only your birth date or

your provider-assigned number?

- Are you aware that if you attend a health fair or a medical screening or participate in email lists, the health information you provide may not be covered by HIPAA protections or other federal and state privacy rules?
- Do you periodically ask your providers for a copy of your health record so you can review it for services provided to you and make sure they agree with your own recollection?
- Have you talked with your pharmacist about the release of your prescription information to marketing organizations and made it clear to them your desires?
- Do you keep a close check on your health insurance card, treating it like a credit card and reporting immediately if it is stolen so someone else cannot make claims against it?
- Do you carefully read the material sent to you by your health plan so that you can be sure that the benefit's statement matches the services you are receiving?
- Are there opt-in and opt-out permissions that a RHIO or health information exchange allows you for the use of your data and the flow of your information among the participants in the RHIO or HIE?

Key Points

1. When you seek healthcare from a provider, clinic, or hospital, your health information is seen by so many healthcare professionals, individuals who work for your health plan, pharmacists, therapists, and other healthcare workers that there is a real danger that unauthorized individuals can access your information in spite of the laws and technology intended to prevent this from happening.

2. There are laws and regulations (HIPAA and the Privacy Rule) that establish standards for the use and dissemination of your health information, protect you, and provide you with the right to access your health information. This legislation also defines the criminal and civil penalties for abuses.

3. The American Recovery and Reinvestment Act of 2009, ARRA, tightened the HIPPA and Privacy Rule and extended coverage of these

regulations to "business associates" who work with your health plan or providers to offer you additional protection.

4. State and case law further protect your health information from penetration by unauthorized individuals or groups and tend to include stricter privacy statutes than those issued by the federal government. Each state is different, so it behooves you to check out the laws specific to where you reside.

5. More than one in four Americans report in surveys that they are concerned with privacy. It is not a trivial matter.

6. Online storage of medical and personal health records and the large databases that are amassed online as a result are inherently dangerous to the integrity of your information and the possibility that aspects of your health information could fall into the wrong hands.

Notes

1. R. Mishra, "Confidential Medical Records Are Not Always Private," The Boston Globe, August 1, 2000.

2. "Privacy Brief, Summary of the HIPAA Privacy Rule State Regulations," The United States Department of Health and Human Services, Office for Civil Rights, April 2003, p. 11.

3. American Recovery and Reinvestment Act of 2009 Subtitle C Section 13405(d).

4. www.cms.gov/qualityinitiativesgeneralinfo

5. Stenberg v. Carhart, 530 US 914, 934–38 (2000); Carhart v. Gonzales, 331 F. Supp. 2d 805, 923–29, 1012–18(D. Neb. 2004); National Abortion Federation v. Gonzales, 330 F. Supp. 2d 436, 470–82 (S.D.N.Y. 2004); Planned Parenthood Federation of America v. Gonzales, 320 F. Supp. 2d 957, 1028–29, 1033–34 (N.D. Cal. 2004).

6. Weld and Kelly vs. CVS Pharmacy, Inc., et al: Civil Action No. 98-0897-F, Superior Court Department of the Trial Court, Commonwealth of Massachusetts

7. Whalen v. Roe, 429 US 589, 1977,

8. http://biotech.law.Isu.edu/cases/ reporting/whalen.html

CHAPTER TEN
Managing Healthcare Costs

The current healthcare market does not work at all like other markets. Unlike cars, computers, or retailing, for example, healthcare services have not become better and cheaper over time. Instead, they have become more costly and people worry about quality too. Why? Because of gross distortions in supply and demand. Consumers, the demanders, have little of the information that interests them, cannot express their feelings about the price because they rarely see the real cost of their health insurance or healthcare purchases, and have an artificially constrained range of choices.

Regina, Herzlinger, *Who Killed Healthcare?*, McGraw Hill p.145, 2007

Ben Green received a letter from a collection agency for services rendered while his wife was in the hospital recuperating from surgery. Unknown to Ben, his employer's health insurance company refused to pay the bill for several diagnostic tests that his wife needed while hospitalized. At the beginning of the year, when Ben renewed his policy, offered through his employer, Ben thought he had opted for the same program he had enrolled in the previous year. However, the coverage terms had changed. Ben, who was traveling on business when the company held meetings to review policy options, had neglected to read the fine print when the policy came in the mail.

Ben's story is not unique; it is typical. Most people spend more time weighing and measuring features of a cell phone, an automobile, or a suit before making a purchase, than they do considering which health insurance plan to sign up for. When they suffer a healthcare crisis and find out the details of their coverage after receiving hospital bills that they cannot pay, it is too late.

Part of the health spending problem in American medicine lies in our inherent belief that healthcare is a right and not a privilege. Americans feel they should receive not only basic care but the best care that medical science can provide, no matter how they may abuse their personal health. Although most people contribute something to the cost of their care, they expect the rest to come from their insurer, the government, or some other source. Despite increased demand for patient care in 2021, the growth in national health spending is estimated to have slowed to 4.2%, from 9.7% in 2020, as supplemental funding for public health activity and other federal programs, specifically those associated with the COVID-19 pandemic, declined significantly. However, the good news is that annual growth in national health spending is expected to average 5.1% over the years 2021–2030, and to reach nearly $6.8 trillion by 2030. Growth in the nation's Gross Domestic Product (GDP) is also projected to be 5.1% annually over the same period. As a result, the comparable projected rates of growth, the health share of GDP is expected to be 19.6% in 2030, nearly the same as the 2020 share of 19.7%. (1).

In his book, *Crisis of Abundance: Rethinking How We Pay for Healthcare*, economist and author Arnold Kling writes that Americans expect "premium medicine." Kling contends that Americans routinely expect to be allowed to see medical specialists and avail themselves of advanced technology. They expect to undergo screening procedures such as MRIs, CT scans, mammograms, and colonoscopies routinely, and want heroic measures during the last ninety days of life when the largest sums of healthcare dollars are spent. Procedures such as angioplasty and heart bypass surgery are done far more frequently in the United States than in European countries, yet there is no hard data that indicates that these procedures positively affect outcomes. (2)

A quick snapshot using data from the CDC National Center for Health shows that in spite of the huge amounts of money spent on healthcare in the U.S., the outcomes are poor. Studies show the following:

- Commonwealth Fund compared the quality of the U.S. system with five other countries found that despite spending twice as much per capita, the United States ranks last or near last on basic performance measures of quality, access, efficiency, equity, and healthy lives. "The United States stands out as the only nation that does not assure access to healthcare through universal coverage."

- A Robert Wood Johnson Foundation study comparing the United States and ten European countries found that the United States had a much higher prevalence of nine of ten conditions, including cancer, heart disease, and stroke, in its population over age fifty.

- More than 70% of the deaths in the United States result from behavioral and lifestyle choices such as diet, exercise, smoking, and obesity, which are primary contributors to the leading causes of death in the United States.

- More than forty million adults stated that they needed but did not receive one or more of essential health services, including: medical care, prescription medicines, mental healthcare, dental care, or eyeglasses, because they could not afford the cost of these services.

- For decades, the United States has been slipping in international rankings of life expectancy, as other countries improve healthcare, nutrition, and lifestyles. Life expectancy in the U.S. ranks forty-second, down from eleventh two decades earlier. Several factors have contributed to our falling behind other industrialized nations, including the fact that forty-five million Americans lack any health insurance, while almost every developed nation in the world offers universal healthcare. (3)

Your Healthcare Choices

Mary was an accountant with one of the large national accounting firms. She loved her job and never took a sick day in the thirteen years she was with the company. She had not even bothered to find a doctor who accepted her insurance plan. At age thirty-five, Mary was diagnosed with lupus, a disease in which the immune system attacks healthy tissue. Her disease progressed rapidly, and she was unable to work and lost her job. Several visits to her doctors, a seizure which sent her to the emergency room, high doses of expensive medications, and a requirement for CAT scans every six weeks to monitor her disease resulted in mounting bills that her insurance company began to question and eventually stopped paying. Mary was a fiercely proud woman. When the payments stopped, she stopped seeing her doctors and stopped getting tests that would monitor her disease and prevent even more insidious events. She turned to the government for assistance, but the state Medicaid bureaucracy was merely another maze to wander through. She became depressed and refused to seek further assistance. Two years after her diagnosis, Mary died of complications.

Although lupus can be fatal, most people who have it live a normal life span. The disease is usually managed with appropriate medication. However, when insurance companies refuse to pay for healthcare costs and individuals deplete their own resources, the medications and the treatments stop, and productive citizens are not able to thrive. When you or family members are hit with a health crisis, you are might be outraged at the bills you receive that you have not authorized, but are expected to pay. Fueling your anger and puzzlement is the fact that your friend in another state had the same procedure you had and paid a lot less than you.

To make matters worse, a frightening occurrence in October 2019, publicized by ProPublica, (a nonprofit online news organization that investigates abuses of power), reported that 30 patients from Coffeyille, Kansas were hauled into court by debt collectors after being served with subpoenas that were issued because they had unpaid medical bills owed to local hospitals. They were presented with warrants for their arrest. Some of these patients were able to find a way to pay their debt. Eleven people were jailed, in spite of the fact that they were ill (4).

The issue here is a question of how an average, hardworking individual can get into medical debt. What are the decisions you must make to ensure that the health coverage you choose is sufficient in the event of a medical emergency? This is also why, when making an employment decision, health benefits have become one of the most important factors, second only to wages.

The majority of people in the United. States do not t buy their own health insurance, they get their insurance through an employer. This is referred to as group insurance or employer-sponsored insurance. According to a 2017 Kaiser Family Foundation (KFF) survey, almost half (49%) of people have employer-sponsored insurance and under 7% of people buy individual health insurance. Employers negotiate group policies that are typically considerably less expensive than individual plans. Some employers offer only one health insurance option, but most give you a choice of plans (i.e., fee-for-service plan, HMO, POS, or PPO). As the cost of health insurance has increased, the number of choices employers offer has decreased, and the healthcare burden grows more cumbersome for everyone. Usually when you choose a health plan, you must keep it for the year. During the open enrollment period at the end of the year, you can choose another option if it is offered

Jeanie is a financial analyst who has been working for one of the large banking institutions for ten years. Her husband is an independent consultant and works for himself. They have three children. The family's health coverage is provided through Jeanie's employer. Jeanie has often thought about starting her own business, but the bank provides comprehensive health insurance. As an entrepreneur, she would have to find coverage elsewhere. Since her daughter has asthma and her son requires frequent checkups for junior rheumatoid arthritis, Jeanie is unhappily locked into her job because she cannot risk the loss of this health coverage for herself and her family.

If you or your family member leaves the job that provides your health insurance, you will lose your employer-supported group coverage. A federal law, the Consolidated Omnibus Budget Reconciliation Act of 1985 (COBRA), makes it possible for most people to continue their group health coverage for a period of eighteen months after leaving a job. You pay the full premium at the group rate.

Some employers do not offer health insurance. There are alternative sources, including: membership in a union, professional associations, and other organizations. Individual insurance is available under fee-for-service, HMO, and PPO arrangements, but the coverage and the costs vary and are often prohibitive. There is also insurance available to individuals through the Affordable Care Act passed in March 2010, often called Obamacare. Before buying any health insurance policy, you need to be sure that you know what it will pay for and what it will not. The following tips should help you

- Shop carefully; policies differ widely in coverage and cost.
- Make sure the policy protects you from large medical costs.
- Read and understand the policy, including the fine print. You do not want surprises when you are sick or in the hospital.

- Check to see when the policy will begin paying and what is covered or excluded.
- Make sure there is a "free look" clause—a period to look over the policy after you receive it to decide whether or not it is right for you.
- Beware of single-disease insurance policies that offer protection only for one disease, such as cancer. A basic policy should cover everything, so you do not need these additional protections, which can be expensive.
- Ask if the plan provides the specific benefits and services you need? Are those services available quickly and efficiently?
- Can you go to the physicians and hospitals best situated for you?
- How easy is it to see a specialist?
- How is the plan ranked against its competitors and by its subscribers?
- Do members get the therapy treatments they need?
- Does the plan provide preventive or wellness services?
- Is the plan accredited?
- How easy is it to see a specialist?
- How is the plan ranked against its competitors and by its subscribers?
- Do members get the therapy treatments they need?
- Does the plan provide preventive or wellness services?
- Is the plan accredited?

Options for a Health Plan?

There are five configurations of health plans for those who do not receive Medicare or Medicaid. The following guidelines provide you with the pros and cons of each type of plan.

Indemnity Insurance (also called fee-for-service)

Indemnity plans allow you to direct your own healthcare and visit almost any doctor or hospital you like. The insurance company pays a set portion of your total charges. Indemnity plans are also referred to as "fee-for-service" plans Under an Indemnity plan, you may see whatever doctors or specialists you like, with no referrals required. Though you may choose to get the majority of your basic care from a single doctor, your insurance company will not require you to choose a primary care physician. An Indemnity plan may also require that you pay up front for services and then submit a claim to the insurance company for reimbursement.

You'll likely be required to pay an annual deductible before the insurance company begins to pay on your claims. Once your deductible has been met, the insurance company will typically pay your claims at a set percentage of the "usual, customary and reasonable (UCR) rate" for the service. The UCR rate is the amount that healthcare providers in your area typically charge for any given service. Although Indemnity insurance coverage pays for most of your health problems it does not typically pay for preventive care like well child visits and annual physical exams, and does not generally cover the total cost of your healthcare. You pay the difference with your co-pay.

Questions to Ask About Indemnity/Fee-for-Service Insurance

1. How much is the monthly premium, and what does that total for a year?
2. What does the policy cover for specific health events? Does it include prescription drug, out-of-hospital care, rehabilitation, or home care? Does it include lab fees and emergency room visits? Do you have the option to see a specialist, and how much will that cost out of pocket?
3. Are there limits on the number of days insurance will pay for hospital or rehab services?
4. Are there some medical conditions that are not be covered by the plan?
5. How do you resolve treatment limit conflicts?
6. Are there waiting periods involved with coverage?
7. What is the deductible? Can you lower your monthly premiums by raising the deductible?
8. Is there a maximum that you would pay out of pocket per year?
9. Is there a lifetime maximum cap that the insurance will pay?
10. How are Indemnity plans rated for customer service?

Preferred Provider Organization (PPO)

The PPO is a form of managed care closest to an indemnity plan. A PPO has arrangements with doctors, hospitals, and other providers who have agreed to accept lower fees from the insurer for their services. As a result, your cost sharing is typically lower within the network (the approved doctors in the plan) but considerably higher if you go out of the network. The PPO always includes a network of providers who have agreed to provide care to patients, subject to contractually established reimbursement levels. The PPO covers many of your healthcare needs for a small per-visit fee (co-payment) as long as you choose from the list of "preferred providers." You can choose to see a doctor who is not on the list, but you will pay the difference between what the provider charges and what the plan will pay, which could be significant. PPOs are the most popular form of health insurance in the United States, with over 148 million Americans participating in these arrangements.

Questions to Ask about a PPO

What is the total annual cost of the PPO, including premiums, co-payments, deductibles, and co-insurance care for doctors outside the PPO network?

1. Who are the doctors in the PPO network? Where are they located?
2. Are they accepting new patients?
3. How are referrals to specialists handled?
4. What hospitals are available through the PPO?
5. How is emergency care handled?
6. What is their customer service like?
7. Can you max out your out-of-pocket deductibles?

Health Maintenance Organization (HMO)

An HMO gives you access to certain doctors and hospitals within its network. A network is made up of providers who have agreed to lower their rates for plan members and also meet quality standards. Care under an HMO plan is covered only if you see a provider within that HMO's network. There are few opportunities to see a non-network provider. There are also typically more restrictions on coverage than other plans, such as allowing only a certain number of visits, tests or treatments during the duration of the plan, including preventive care, for a set monthly fee. There are many types of HMOs. In some, the doctors are employees of the health plan, and you visit them at a central location. Other HMOs contract with physician groups or individual doctors who you will see in their private offices. Every HMO will give you a list of doctors from which you choose a primary care physician. That doctor coordinates your care and handles referrals when you need to see a specialist. With some HMOs, you pay nothing for an office visit, while others require a co-payment. Most HMOs cover all of your healthcare needs, including checkups, immunizations, and hospitalizations, often with a co-payment. If you are forced to go "out of network" for a particular specialist that is not listed, you need to request permission, and you may have to contribute a large percentage of the charges for this service.

Features of HMOs

- HMO plans may require you to select a primary care physician (PCP), who will determine what treatment you need.
- With some plans, you may need a PCP referral so you are covered when you see a specialist or have a special test done.
- If you choose to see a doctor outside of an HMO network, the plan will not cover that visit. and you pay the entire cost of medical services.
- Premiums are generally lower for HMO plans, and there is usually no deductible or a low one.

Questions to Ask

1. What will the HMO plan cost each year, including co-payments?
2. Are there many doctors to choose from? Which doctors are accepting new patients? How hard is it to change doctors?
3. How are referrals to specialists handled? What is the payment policy if a specialist is outside the HMO network?
4. Is it easy to get appointments with doctors in the HMO? How far in advance must routine visits be scheduled?
5. What arrangements does the HMO have for handling emergency care?
6. What hospitals are in the network?
7. Are there limits on services such as preventive medical tests, surgery, mental health, and home care?
8. What will the HMO plan cost each year, including co-payments?
9. Are there many doctors to choose from? Which doctors are accepting new patients? How hard is it to change doctors?
10. How are referrals to specialists handled? What is the payment policy if a specialist is outside the HMO network?
11. Is it easy to get appointments with doctors in the HMO? How far in advance must routine visits be scheduled?
12. What arrangements does the HMO have for handling emergency care?

13. What hospitals are in the network?
14. Are there limits on services such as preventive medical tests, surgery, mental health, and home care?

Point of Service

Many HMOs offer an indemnity type option known as a Point of Service Plan. The POS is basic, managed care with lower medical costs in exchange for more limited choice. When you enroll in a POS, you choose a primary care physician who must be from within the healthcare network. Unlike an HMO, your POS physician may make referrals outside the network. You will pay for these services, but they are available. The advantages of a POS are that you have a much broader choice of doctors and your annual out-of-pocket costs are limited. A disadvantage is that your co-payments for non-network care are high. Overall, the costs will be slightly lower than a PPO because most of your healthcare is controlled by your managed care insurer.

Questions to Ask about POS Health Insurance

1. How many doctors are there to choose from? Where are they located?
2. Are doctors in the network private or group practice?
3. How are referrals to specialists handled?
4. What hospitals are in the plan?
5. What services are covered and are there limits to treatment?
6. How much is the annual premium?
7. What is the deductible for non-network care? Is there an out-of-pocket maximum?
8. What arrangements are provided for services that are not offered by the HMO? Who covers the payments?

The Patient Protection and Affordable Care Act (PPACA)

When Dan was laid off from his engineering job, he lost not only his work, his income, and his feeling of self-worth, but his health insurance as well. COBRA payments were out of the question, as Dan did not have the resources to keep up with the premiums.

Diagnosed with depression six months after the layoff and unable to find other work, Dan also lost his condo, his car, and his credit cards, and he ended up homeless. Having had asthma all of his life, Dan developed breathing problems and contracted pneumonia. He landed in the emergency room of a large urban public hospital and was admitted.

Dan became one of nearly fifty million uninsured American individuals who should have some recourse for finding healthcare before an illness precipitate into a major crisis that risks his life and sends healthcare costs for everyone spiraling upward. In a bad economy, this situation could happen to anyone.

PPACA, signed into law on March 23, 2010, has been the subject of much heated discussion, political posturing, and debate. A decade later, it is still clear that U.S. healthcare expenditures that as of 2020, accounted for 19.7% of GDP, and yet that high sum was not buying all US citizens the kind of healthy life that is enjoyed in other countries. On the other hand, the PPACA represents an attempt to address many deficiencies in US healthcare and makes it possible for more people to have healthcare coverage when they cannot get it through their employment. Here are a few highlights of PPACA that you need to understand:

1. Beginning in 2011, health insurance companies were required to spend at least 80% of the premium dollars they collect on direct medical care and quality improvement activities. Furthermore, they must publicly report how they spend those dollars. This is intended to reduce health disparities, bring down healthcare costs, increase investment in prevention and wellness, to give you and your family more control over your care.

2. Starting with plan years beginning on or after September 23, 2010, the law requires health plans and health insurance policies to cover certain recommended preventive services at no charge to patients in an effort to encourage you to live a healthier life. This includes:
   a. Medicare patients will be eligible for an annual wellness exam and certain preventive services with no cost-sharing.
   b. States and local communities will be given new resources to address chronic diseases, obesity, smoking, and drug use.

c. Additional provisions enable consumers who have joined new health plans to receive cost-free preventive services like regular check-ups, cancer screenings, and immunizations at no additional cost; allow citizens the choice of a primary care doctor, ob./ gyn, and pediatrician; and enable them to use the closest emergency room without penalty

3. The act has a provision that enables you to keep young adults on a parent's plan until age twenty-six.

4. It includes a mandate that insurers will no longer be able to deny coverage to children with preexisting conditions, places lifetime limits on benefits, ensures that plans cannot cancel a policy retroactively without proving fraud, and deny claims without the chance for appeal.

5. The law forbids insurers from refusing to cover you on the basis of preexisting conditions and prevents your health plan from assessing you with a higher co-payment or co-insurance for out-of-network emergency room services.

6. The law bans insurance discrimination so people who have been sick can't be excluded from coverage or charged higher premiums, including women who no longer have to pay higher premiums because of gender.

7. The Affordable Care Act mandates a new health insurance marketplace to be established by 2014 consisting of health insurance exchanges that offer one-stop shopping so individuals can compare prices, benefits, and health plan performance on easy-to-use websites. The exchanges guarantee that all people have a choice for quality and affordable health insurance even if a job loss, job switch, move, or illness occurs. The new law also provides tax credits to help Americans pay for insurance.

8. The law expands initiatives to increase racial and ethnic diversity in the healthcare professions. It strengthens cultural competency training for all healthcare providers and requires health plans to use language services and community outreach in underserved communities.

9. The act expands the healthcare workforce and increases funding for community health centers, including health teams that can manage chronic diseases and comprehensive health services for everyone, no matter how much they are able to pay.

10 The PPACA law provides new funds for home visits for expectant mothers and newborns in an effort to reduce infant mortality and post-birth complications.

11. New rules included in the Health Reform Act will simplify paperwork for physicians and lessen their administrative hassle, presuming that they have made investments in electronic health records, to help doctors and hospitals focus more on patient care and less on paperwork.

12. Investments in new models of patient-centered, coordinated care, including investments in medical homes and other advanced care coordination and disease management tools, are included to give patients more control over how your care is delivered and enables your providers to be rewarded with bonuses when they meet certain criteria.

13. PPACA supports the training and development of more than sixteen thousand new primary care providers over the next five years and establishes new nurse-managed health clinics to train nurse practitioners to work in underserved communities. It also provides the structure for The National Health Service Corps, a program that repays loans and gives scholarships to primary care providers who work in areas of the country with too few health professionals.

14. Starting with policies issued or renewed on or after September 23, 2010, insurance companies will be prohibited from dropping patients from coverage when they get sick because of an unintentional mistake on a form. By 2014, annual limits are phased out that adversely affect the sickest patients with the highest costs.

15. Provisions that went into effect on September 23, 2010, address the Medicare drug coverage gap, also called the "donut hole." This means that after you and your plan have spent a certain amount of money for covered drugs, you have to pay all costs out-of-pocket for your drugs (up to a limit). Starting in 2011, if you have high prescription drug costs that put you in the donut hole, you'll get a 50% discount on covered brand-name drugs while you're in the donut hole. Between 2010 and 2020, you'll get continuous Medicare coverage for your prescription drugs.

16. The new law extends the life of the Medicare Trust fund at least twelve years and takes other actions to secure the program.

17. The legislation does not make any changes in COBRA. Coupled with that, the Early Retiree Reinsurance Program (ERRP) provides much-needed financial relief to businesses, schools, and other educational institutions, unions, state and local governments, and nonprofits in order to help retirees and their families continue to have quality, affordable health coverage.

Other Elements of the ACA that you should know:

The ACA requires health plans that offer health insurance coverage to individuals or in the small group market to ensure that such coverage includes the essential health benefits package and specifies that a group health plan ensure that any annual cost-sharing imposed under the plan does not exceed specified limitations.

The ACA prohibits a health plan from applying any waiting period for coverage that exceeds 90 days; or discriminating against individual participation in clinical trials. The legislation does not make any changes to COBRA benefits which mandates an insurance program give employees the ability to continue with their health plan for a specified number of months after leaving their place of employment. The law provides initiatives to increase racial and ethnic diversity in the health professions and strengthens cultural competency training for all providers. It requires health plans to use language services and community outreach in under-served communities.

The ACA extends the life of the Medicare Trust fund at least 12 years,

The ACA requires each US hospital to establish and make public a list of its standard charges for items and services.

The ACA requires health plans to implement an effective process for appeals of coverage determination and claims.

Evaluating the Right Plan

The health insurance exchange system provides a sliding scale of credits for low- and moderate-income individuals and families, based on your most recent tax return. The credits are most generous for those who are just above the proposed Medicaid eligibility levels and decline as your income goes up. They are completely phased out when your income reaches 400% of the federal poverty level ($43,000 for an individual or $88.000 for a family of four. The maximum out-of-pocket costs for any Marketplace plan for 2014 are $6,350 for an individual plan and $12,700 for a family plan.

The Kaiser Family Foundation has developed a calculator to help people determine whether they are eligible for a subsidy and how much. This tool helps individual who are purchasing insurance directly on the exchanges figure out their premiums and eligibility for subsidies. With this calculator, you can enter different income levels, ages, and family sizes to get an estimate of your eligibility for subsidies and how much you could spend on health insurance.

You can use this calculator at: http://kff.org/interactive/subsidy-calculator/

There are four plans and a Catastrophic Plan, offered by the insurance exchanges: Platinum, Gold Silver, and Bronze. The choice of which plan best suits you is essentially a decision of whether you can afford to pay more up front or pay when you receive care. All qualified health plans cover essential health benefits. A qualified health plan is certified by an exchange, provides essential health benefits, provides coverage at a "metal level" of actuarial value (bronze [60%], silver [70%], gold [80%], or platinum [90%]) follows established limits on cost-sharing, such as deductibles, copayments, and out-of-pocket maximum amounts, and meets certain other requirements determined in the ACA, or by state exchanges.

It is important to note that many of the health plans sold on the insurance exchanges offer substantially fewer choices of hospitals and physicians than current health plans have offered in the past. Catastrophic plans are available to people under 30 years old and to people who have hardship exemptions from the fee that most people without health coverage must pay. They have very high deductibles and essentially provide protection for worst-case scenarios, like a serious accident or extended illness.

All of the plans offer essential health benefits including:

- Ambulatory patient services (outpatient care you get without being admitted to a hospital)
- Emergency services
- Hospitalization

- Maternity and newborn care (care before and after your baby is born)
- Mental health and substance use disorder services, including behavioral health treatment
- Prescription drugs
- Rehabilitative and facilitative services and devices (services and devices to help people with injuries, disabilities, or chronic conditions gain or recover mental and physical skills)
- Laboratory services
- Preventive and wellness services and chronic disease management
- Pediatric services, including dental and vision care

There are two choices that you have to make when you go to your state health insurance exchange:

1. Which plan best suits you Bronze, Silver Gold Platinum, and Catastrophic?
2. Who will be your insurance carrier?

The platinum plan has several levels of premium. For example, one platinum plan might have a high $1000 deductible paired with a low 5% coinsurance. A competing platinum plan might have a lower $400 deductible paired with a higher coinsurance and a $10 copay for prescriptions. You would pick a platinum plan based on whether or not you can afford the higher premiums each month.

The percentages of healthcare costs you pay for each type of plan are as follows:

Platinum plan:	10%
Silver plan:	30%
Gold plan:	20%
Bronze plan:	40%

Questions to Ask When Choosing a Plan Type or Carrier (e.g., Aetna, Universal Health, Blue Cross/Shield, etc.)

1. What will the plan cost per year, including deductibles and co-payments?
2. Who are the doctors in the network? Where are they located? Are they accepting new patients?

3. How are referrals to specialists handled?
4. How well does the plan do on quality and patient satisfaction measures?
5. What hospitals are available to me?
6. How is emergency care handled?
7. Do any of us have frequent hospitalizations or chronic conditions?
8. Are there limits on services such as immunizations, medical procedures, surgeries, mental health and home care?
9. Would I benefit from having a Flexible Spending Account where I set aside a portion of my earnings to pay for medical expenses or dependent care?
10. Do I or members of my family expect a lot of doctor visits in the next year?
11. Do I or members of my family require ongoing medications? Are they covered by the plan?
12. **Is there an appeals process if I go to my current doctor and find out later that my new plan doesn't cover that practice?**

How to Apply for Insurance through the Exchanges:

Most states have their own exchange. When you fill out an application, you can compare plans side-by-side, based on price and other features important to you. You will also learn if you can save money on your monthly premiums or get lower out-of-pocket costs.

The ACA requires that each health insurance exchange have two certified navigators, one of whom must be not-for-profit. The navigators must provide "fair, impartial, and accurate information to assist you with submitting the eligibility application, clarifying distinctions about [qualified health plans], and helping qualified individuals make informed decisions during the health plan selection process. They also provide assistance to consumers who are disabled, do not speak English, or who are unfamiliar with health insurance. They will ensure that your plan includes an annual wellness visit and help you with issues of chronic disease and disability. Additionally, each state has a Department of Insurance where there are individuals who should be able to assist you with questions and problems related to helping you understand the ACA and your rights and benefits.

At the Health Insurance web site, will find an Application for Health Coverage and Help Paying Costs. It is best if you fill out the application online, although you can submit a paper application or call your Marketplace call center and apply over the phone. To complete the Application for Health Coverage, you will need to create a secure personal account with a login ID and password.

Tips for online applicants:

- Use laptop or desktop computer instead of an iPad or smartphone.
- Use the best browser for your computer's operating system.
- Clear your old cookies and clear your cache (stored URLs you have recently visited).
- Make sure your browser is set up to accept cookies.
- You may not be able to complete an online application with a mobile device, like a smartphone or tablet.
- Go to https://www.healthcare.gov/marketplace/individual/.
- Select your state then follow the directions for filling out the application.

The Application asks you basic information about yourself (and any family members who are applying for coverage with you) including your Social Security number and information about your citizenship or immigration status. It also asks employment and income information, including information from your most recent income tax return. Once you've submitted the application, you will make a choice among the qualified health plans in that exchange and will receive notice of whether are eligible for help paying for the plan.

Individuals who choose not to obtain coverage will pay a penalty of 2.5% of their modified adjusted gross income above a specified level. Employers have the option of providing health insurance coverage for their workers or contributing funds on their behalf. If they fail to do that they will be penalized.

Beginning in 2014, businesses with 50 or fewer full-time equivalent (FTE) employees, can use the Small Business Health Options Program (SHOP) to offer coverage plans that fit their needs and budget. The advantage of using SHOP is that you control the coverage you offer and how much you pay toward employee premiums. Small businesses also choose from the four levels of coverage for a single plan that meets the needs of your business and employees. Business with fewer than 25 employees are not required to provide insurance to their employees. If they choose to offer coverage through SHOP, they may qualify for a small business healthcare tax credit worth up to 50% of the premium costs. If you are self-employed with no employees, you can get coverage through the individual market exchanges, but not through SHOP.

Checklist

- Check with your insurance company to make sure you are officially enrolled.
- Be sure to pay your first premium so that your plan will kick in.
- Carefully review your member card or other materials your plan sends you. All health plans must provide consumers with a Summary of Benefits and Coverage (SBC). This is a brief, clearly written description of what your plan covers and how it works. The SBC is posted for each plan on the Marketplace web site and makes it easier for you to compare the differences in health plan benefits and cost sharing. If you didn't get a card or if there are errors in your SBC, contact your insurance company.
- Review your plan's provider directory (they are online) to locate a doctor who is in your particular plan. Your insurance plan should have a list of primary care physicians (PCP) available to you in your particular plan.
- Make an appointment with the PCP you have chosen. The doctor will want to examine you, see your previous medical records, your medications, get your past history, labs, imaging studies, etc. All plans have a deductible which must be paid by you before your plan covers your cost of care. It is important to know what that deductible amount is.
- Check with your insurance company to see which pharmacies accept your plan and choose one that is convenient for you. Your new PCP needs to know which pharmacy you'll be using. Make sure that the pharmacy you select is in network with your insurance plan. Also be sure that should you need other services such as imaging or PT those providers are also in your network so you will not have to pay a significant out-of-network co-payment.

- Pay special attention to the out-of-pocket maximum. You want to insure that, if you were to have a catastrophic illness, that you are able to cover the maximum amount before insurance would pick up the rest.

- Be sure you check your health plan's prescription drug "formulary" and other online information about the plan. The "formulary" is a list of prescription drugs the plan will cover. If you don't find your medication on the formulary but your doctor says it's medically necessary for you to take that specific drug, you can appeal for an exception to the plan formulary. If there is a Consumer Assistance Program in your state, there are individuals in this program who can help you file an appeal.

- Always check to see if a drug you have been prescribed and need is available in a generic form.

- Pay attention to requirements such as Proof of Coverage - When you file your 2014 tax return (most people will do this by April 15, 2015) you will have to enter information about your coverage (or your exemption) on the return. You should get a notice from your insurance provider by January 31, 2015, describing your coverage status during the previous year.

- Make an effort to keep up with information regarding your coverage at www.healthcare.gov. You can log in to your account at any time, and click on your application. You will see a summary of your coverage on the "My Coverage" page. Click the links for more details about your plan benefits.

- If you want health insurance but cannot afford it, you may be eligible for Medicaid. The Federal government has attempted to expand the Medicaid program but not all states are complying with the request. The United States Supreme Court, ruled, in June 2012, that states cannot be forced to make that change. In those states that offer expanded Medicaid, anyone with an income at or lower than 138% of the federal poverty level, (about $16,000 for an individual or $32,500 for a family of four based on current guidelines) will be eligible.

- Learn about the appeals process. If your health insurance company doesn't pay for a visit to the doctor, you have the right to appeal the decision and have it reviewed by an independent third party.

- Ask about alternate facilities to receive care within the network. Doctors often work at outpatient surgery centers as well as hospitals and what they charge can vary by location.

- Find lower-cost after hour care at urgent care centers and retail clinics if you can. The cost differential between those facilities and the emergency room can amount to hundreds of dollars. Check to see if your insurance company will waive the co-pay when you go to a convenient, economical care center and if they increase your co-payment if you go to the ER. Although there are times when you have to go to the ER it is best to know your options before a need occurs.
- Ask about discounts. Some plans offer additional discounts as an incentive to use certain providers.
- Take advantage of your wellness benefits so that you have a better chance of remaining healthy. Your wellness and fitness are in your hands and no one else can do it for you.
- Request help for chronic conditions. Specifically, ask your primary care doctor to set you up with a way to monitor your conditions without frequent visits to the office or the hospital so that you avoid a crisis with prevention.
- (Note that large employers cannot send their employees to state insurance exchanges for their health insurance)

Medicare

Medicare is the federal health insurance program for American ages sixty-five and older and for certain disabled Americans. If you are eligible for social security or railroad retirement benefits and are age sixty-five, you automatically qualify for Medicare, provided that you paid the Medicare payroll tax for at least ten years of your work life and are a citizen or permanent resident of the United States.

Medicare has three parts:

Part A helps you pay for hospital care and some other care such as home health, hospice, and skilled nursing facility care. Medicare Part A does not require you to pay a monthly fee or premium.

Part B helps pay for doctors' visits, some home healthcare, medical equipment, some preventive services, outpatient hospital care, physical therapy, laboratory tests, X-rays, mental health services, ambulance services and blood. With Medicare Part B, you pay a premium and have the option of choosing which coverage plan is best for you.

Part D, the prescription drug plan, subsidizes the costs of prescription drugs for Medicare beneficiaries. Under the Medicare Modernization Act, as of January 1, 2006, insurance companies and other private companies are mandated to work with Medicare to offer Part D drug plans and determine discounts on drug prices. These drug plans vary in terms of what prescription drugs are covered, how much you have to pay, and which pharmacies you are allowed to use. People with an income at or below a set amount and with limited resources will qualify for extra help under Medicare Part D. This help is intended to assist you with a drug's monthly premium and with costs you would normally have to pay for prescriptions.

Medicare includes two approaches that you may choose from:

1. **The Original Medicare Plan** is a traditional fee-for-service plan where you can choose any doctor or hospital you want and Medicare will pay a share of the costs while you pay the rest.
2. **Medicare Managed Care Plans** offered by private insurance companies that pay for the same services as the Original Medicare Plan as well as for additional healthcare services generally with a co-payment every time you visit the doctor, hospital, or use other healthcare services. These plans are available in most but not all parts of the country.

In other words, Medicare is a subsidized service, not a free service. How much you pay depends on which plan you choose, how often you go to the doctor or hospital, whether you have other health insurance, and whether you qualify for help through other state and publicly supported health programs.

Medigap

You may want to consider purchasing a private insurance policy called medigap to pay medical bills that Medicare does not cover. Medigap policies are sold by private insurance companies to fill the "gaps" in original Medicare plan coverage, and their terms are regulated by federal and state statutes. There are twelve different standardized medigap policies (medigap plans A through L), and their costs vary.

The best information on Medicare is found in the Medicare Handbook, which explains how the Medicare program works and what your benefits are. To order a free copy, write to the Healthcare Financing Administration, Publications N1-26-27, 7500 Security Blvd, Baltimore MD 21244-1850. Or call 1-800-MEDICARE, the twenty-four-hour customer service line. You can also find answers to your questions about Medicare at www.medicare.gov.

Consumer Directed Health Plans (CDHP)

The Bordens are a middle-class American family with two young children, a modest home with a mortgage, a dog, two cars, and a couple of credit cards that always seem to have a monthly balance larger than the Borden's budget. Although both parents work and have received cost of living wage increases over the past several years, there never seems to be enough money to make ends meet. An analysis of their budget reveals that higher payroll deductions from Dad's paycheck for healthcare, which covers the entire family, and higher out-of-pocket expenses for medical needs during those same several years has exceeded their cost-of-living raises.

This same story repeats itself in families across the country. Two hardworking citizens making a genuine contribution to society are being financially buried, many even becoming bankrupt, due to the burden of critical healthcare expenses. Receiving state-of-the-art of healthcare is unachievable for the average American family today.

For years, healthcare spending was strictly the purview of insurers, the government, or your employers. They determined what medical expenses and tests will be covered and what will not. If you opt for a service not covered, you pay for it, in spite of the high premiums you pay each month. This burden of deciding how to spend healthcare dollars has now shifted to the consumer with Consumer Directed Health Plans (CDHPs). These plans were created to provide consumers with more control over their healthcare dollars and expenses. CDHPs are often referred to as three-tier payment systems, consisting of a savings account, out-of-pocket payments, and an insurance plan. As follows:

> **Health Savings Accounts (HSA)**—These are created by individuals who are covered by a qualified health plan that has a high deductible to assist them in paying out-of-plan expenses. Both employees and employers can contribute to an HSA up to an annual limit set by a statutory cap. Employee contributions to an HSA are made on a pre-income tax basis, and some employers' HSA is owned by the individual, who retains the funds if he/she were to leave the job.
>
> **Health Reimbursement Accounts (HRA)**—HRAs are medical care reimbursement plans established by employers and used by employees to pay for healthcare Employers typically commit to a specific amount of money available in the HRA for an individual to pay premiums and other medical expenses. Unspent funds in an HRA are usually carried over to the next year. Employees do not take their HRA balance with them if they leave the job.
>
> **Flexible Spending Account (FSA)**—The flexible spending arrangement allows an employee to set aside a portion of his or her earnings to pay for qualified expenses, most commonly for medical expenses but also for dependent care or other expenses. Money deducted from an employee's pay and put into an FSA is not subject to payroll taxes, resulting in a substantial income tax savings. Paper forms or an FSA debit card, also known as

MY NAME: (as it appears on my Social Security card)					
1. Plan name					
2. Plan Network Location					
3. Type of Plan (ACA only) - *circle one* (Percentage of Cost Coverage)	PLATINUM (90%)	GOLD (80%)	SILVER (70%)	BRONZE (60%)	CATAS-THROPIC (Less than 60%)
4. Health Savings Account (HSA) eligible?	Yes / No				

19. Other places I get care: (ex: residential treatment center, outpatient counseling)			
20. My medication prescriptions:			

	Covered? (circle one)	Do I need a referral or pre-authorization? (circle one)	Co-Pay/Coinsurance?		What are the limits or maximums? (benefit amount or amount of treatments per year)
			In-network	Out-of-network	
21. Primary care visits	Yes / No	Yes / No	$ / %	$ / %	$ /#
22. Specialist visits	Yes / No	Yes / No	$ / %	$ / %	$ /
23. Mental health services	Yes / No	Yes / No	$ / %	$ / %	$ /
24. Emergency room	Yes / No	Yes / No	$ / %	$ / %	$ /
25. Hospital care	Yes / No	Yes / No	$ / %	$ / %	$ /
26. Home health care	Yes / No	Yes / No	$ / %	$ / %	$ /
27. Prescription medicines	Yes / No	Yes / No	$ / %	$ / %	$ /
28. Preventive treatment	Yes / No	Yes / No	$ / %	$ / %	$ /
29. Preventive screenings (ex: depression, substance abuse)	Yes / No	Yes / No	$ / %	$ / %	$ /
30. Other Notes to Self					

12. Is there a separate co-insurance for behavioral/mental health treatment?	Yes / No
13. What is the maximum out-of-pocket expense? (does not include premium)	
Per individual per year:	$
Total family per year:	$
14. Are there any other costs?	$

Section 2: MY HEALTH COVERAGE

	ARE THEY COVERED?	
	In-network	Out-of-network
15. My primary care doctor:		
16. My prescribing provider: (ex: *Psychiatrist, PMHNP*)		
17. My therapy/treatment providers: (ex: *LCSW, Psychologist, Therapist*)		
18. My hospital:		

5. Number of people in my household:	# _____
6. Annual household income:	$
7. Am I eligible for a tax credit?	Yes / No Amount per year: $ _____
8. How much is the premium?	
Amount per month:	$
Amount per year:	$
9. How much is the deductible per year?	$
10. How much are co-pays?	
Primary care visits:	$
Mental health specialist visits:	$
Hospital visits:	$
Emergency room visits:	$
Urgent care visits:	$
Prescription drugs:	$
Other:	$
Other:	$
11. How much is co-insurance?	$_____ or _____%

Key Points

1. US healthcare spending is nearly 20% of the gross domestic product, twice as much as most other countries, with not very much to show for these expenditures when evaluating our mortality rates and other basic quality performance measures.

2. Our personal healthcare costs continue to rise at an alarming rate, with higher premiums and higher out-of-pocket expenses, while our wages have dropped significantly relative to inflation. Therefore, many of you who have health insurance are grossly underinsured in the event of a major catastrophic medical event.

3. To add another element that should not happen in a democratic society, healthcare costs are uneven, with the cost of a procedure in one area of the country or even in one institution within a specific geography vastly different than the cost of the same procedure elsewhere.

4. Individuals have to become significantly better informed about the specifics of their health insurance policies and their options when choosing a health plan.

5. There are good and bad elements in each of the health plans currently available that include: indemnity insurance, managed care with a preferred provider organization, HMOs, and point-of-service plans, the Insurance Exchanges and Medicare/MediGap.

6. For those over the age of sixty-five who are eligible for Medicare, the issues are complex. It is best to seek assistance at: www.medicare.gov, or by ordering a free copy of the Medicare Handbook.

7. Consumer Directed Health Plans enable individuals who get their health insurance through their employers to set aside some healthcare dollars in an HAS, an HRA, or a flexible spending account. This places an added burden on an employee to become an educated consumer/ It also helps you have money set aside for those unexpected medical costs that inevitably add up.

8. The Patient Protection and Affordable Care Act is complex and detailed. There are many elements in the Act that will improve your access to care over time. You need to inform yourself about the details of the PPACA which you can find at www.healthcare.gov.

Notes

1. CMS Office of the Actuary Releases 2021–2030 Projections of National Health Expenditures Mar 28, 2022

2. https://www.cms.gov/newsroom/press-releases/cms-office-actuary-releases-2021-2030-projections-national-health-expenditures

3. Arnold Kling, "The Rise of Premium Medicine," Crisis of Abundance: Rethinking How We Pay for Healthcare, Cato Institute, Washington DC, 2006.

4. Elizabeth Doctor and Robert Berenson, "How Does the Quality of U.S. Healthcare Compare Internationally?" Robert Wood Johnson Foundation, Urban Institute, 2009.

5. ProPublica, October, 2019, https://features.propublica.org/medical-debt/when-medical-debt-collectors-decide-who-gets-arrested-coffeyville-kansas/

6. David Himmelstein, Elizabeth Warren, et al., "Market Watch: Illness and Injury as Contributors to Bankruptcy," Health Affairs Journal, February 2005, doi:10.1377 hlthaff.w5.63.

7. "Faces in the News," Forbes Magazine, Sept. 15, 2005.

Chapter Eleven

Looking Ahead: A Snapshot of your Medical Future

Never before in history has innovation offered promise of so much to so many in so short a time. Bill Gates, Business @ the Speed of Thought: Using a Digital Nervous System with Collins Hemingway, Collins Books, 1999.

Healthcare 2050

Baby James is born on February 12, 2050, at 3:00 A.M. Minutes after his birth, a microchip the size of a grain of rice is embedded in his arm. It is coded with an assigned number that will belong to James for the rest of his life, no matter where in the world he may go, and points to a secure website that was created within hours of James's birth. Stored at this site are all of the tests administered to James in his first few days of life, including the PKU, the APGA, James's blood type and DNA, genetic markers, analysis of James's stem cells and skin tissue, results from a hearing test, a thumb print, and a scan of his retina, as well as his birth certificate and other important information. James (or his family until he becomes of age) will own and control access to this record. It will be available on a need-to-know basis to all his healthcare providers, with consent. This immediate access to James's information eliminates the need for duplicate tests or interventions. It provides the right information for James, no matter where or when he needs care

Imagine the Possibilities:

Imagine a world where every person has a microchip that lasts indefinitely and transmits a unique fifteen-digit number that can be read by a handheld scanner that points to your digital health record every time you see a healthcare provider or go to the emergency department. The doctors who take care of you can add comments and updates to the record to keep it current. You can also enter data in your record.

Imagine a time when electronic devices linked to computers and communications networks allow your implants and medical conditions to be monitored remotely at any time.

Imagine having an operation done by a robotic arm equipped with laser beams that repair, extract, and replace tissue and organs. The procedure is not invasive and enables you to go home from the most intricate surgery hours later. Your doctor is right there overseeing the operation, available to answer your questions and jump in if there is an emergency.

Imagine knowing years before it might happen about a disease you are carrying that is detected by noninvasive brain scans that can also look at the impact of various drugs or chemical imbalances in your body, enabling your doctors to engage in preventive measure that could markedly increase your life span and quality of life.

Imagine having the ability to protect yourself against neurological conditions such as Alzheimer's disease, ALS, Parkinson's disease, or the ravages of cancer by understanding the genetic markers you are born with.

Imagine that when you brush your teeth, your toothbrush examines your gums and takes your blood pressure, blood sugar, and oxygen levels.

Imagine that when you step out of bed, your carpet measures your weight.

Imagine that when you go to the bathroom, your toilet diagnoses your current body chemistry and detects irregularities in your waste.

Imagine that all of this information is collected and transmitted through your smartphone or computer to a remote location where nurses are monitoring you and addressing any irregularities that they detect before they result in a visit to the doctor or the emergency department.

All of these changes and more will be standard in the healthcare you experience in the second half of this century. They will be made possible with sophisticated diagnostic digital tools available to you and to your physicians.

In the next several years, technology will reveal greater insight into the mechanisms of the human body, as well as better education, and incentives that empower you as a patient. This will result in as-yet-unimagined additional discoveries. Communication, convergence, implementation of standards and protocols will make interconnection as transparent as a telephone call or email communications are today. The easy transfer of information will result in a continuous patient-centered care that will become the norm.

Artificial Intelligence (AI) for Devices and Enablers

There are many devices and enabling technologies that were developed in the twentieth century and the first decade of the twenty-first century that will make healthcare 2050 possible. A fundamental start for advancing digital technology in the future is use of AI which combines computer science with complex datasets to enable problem-solving. AI makes it possible to process vast troves of data from radiology films or from hundreds of electronic medical records to identify patterns which enable researchers to predict outcomes and discover new treatments. Examples of such systems include early warnings based on radiological images that help staff spot subtle changes in a patient's condition; and images to identify and develop predictive models that tell clinicians which drugs to test with certain patients for a particular disease; All of this expands decision-making capabilities of the clinicians, who, at the end of the day, are the humans taking care of patients. By using artificial intelligence in the health ecosystem, researchers are now able to use massive amounts of collected data and turn that into meaningful decision-making tools. This will ultimately enable healthcare professionals to deliver care faster, at a lower cost, and with a great deal more accuracy.

Thirty percent of healthcare costs are associated with administrative tasks. According to Business Insider Intelligence, AI automates these tasks, such as pre-authorizing insurance, following-up on unpaid bills. These digital capabilities ease the workload; save money reduce costs; and ultimately provide faster, more efficient coordinated care (1)

The promise of AI in caring for patients will be used in many other ways, going forward, for example, in robotic surgery, in precision medicine; as a virtual assistant (chatbot); and most importantly in analysis of extensive datasets to determine, develop and treat patients in new ways. Using advanced supercomputing algorithms, and deep learning, AI will enhance the cognitive skills of physicians and encourage patients to become more engaged and empowered to manage and monitor their own care.

Chatbots or virtual assistants provide a support system for some patient care. today. In the future, they will fill the gap caused by our shortage of medical care providers (doctors, nurses, nurse practitioners, physician assistants. With built-in artificial intelligence and natural language processing a chatbot can simulate a conversation with a patient through messaging applications, websites, mobile apps. These virtual nursing assistants chatbots help to monitor patients, and answer their questions, in real-time. and enable regular and consistent communication between you and your healthcare providers. These AI-powered virtual assistants provide personalized experiences to patients, helping them detect their illness based on the symptoms, The virtual nursing assistant guides the patient through a course of treatment providing the reassurance and connection that ill patients need. (2)

Radio Frequency Identification (RFID)

Radio Frequency Identification (RFID) is a technology that uses electromagnetic fields to automatically identify and track tags attached to objects. RFID use in healthcare is applied to patients when tiny transmitters embedded in a tag or bracelet send out an encoded stream of bits to alert a reader. During the first decade of the twenty-first century, RFID was used to match an identifier on a patient with an identifier on a pill container or other medication container to check that the right patient was given the right drug at the right dosage. This has helped reduce medical error and promote patient safety in such areas as specimen collection, blood analysis and medication administration, tracking of, patient movement, and tracking medical equipment throughout a hospital. RFID use will increase by 2050, as this technology proves its usefulness in promoting efficiency and patient safety.

Gene Sequencing

The total genome sequencing of DNA is technologically possible today. An invaluable scientific advancement that provides healthcare professionals with insights into disease, treatments and cures today, there are barriers to widespread acceptance including: cost, limitations on how we interpret the results, and unresolved philosophical issues. The pandora's box, however, has been opened, because this science goes to the very core of understanding the workings of the human body. It is a science that will advance so that there will be wide acceptance over time. By 2050, every newborn baby will be sequenced immediately after birth and the knowledge gleaned from that, will result in customized care based on an individual's genetic makeup. Patients in the future will know their risk for a large number of diseases that are tied to their genetic markers. Providers will have vast data at their fingertips that will give them many more options for the treatment of disease. The aggregation of data from large numbers of humans who have been sequenced will form an extensive knowledge base that, with AI/machine learning analysis will foster aggressive disease surveillance and preventive medical treatments. For example, understanding what contributes to longevity or what increases resistance to disease will result in new therapies with widespread population health benefits that will transform medicine.

Precision Medicine

Precision medicine is an emerging approach to the treatment and prevention of disease that takes into account individual characteristics based on genetic makeup, disease biomarkers, treatment history, environment, lifestyle and family. Elements such as poor nutrition, obesity, use of alcohol, and drugs, workplace conditions, living conditions, daily exercise and activity. This approach is in sharp contrast to the one-size-fits- all approach that has been used in medicine for decades to address disease treatment and prevention with less consideration for the differences among individuals. (3)

Precision medicine's foundation relies on the large amounts of data from disruptive innovation, including affordable genome sequencing, advanced biotechnology and patient use of wearable health sensors. Since a human's genes define their risk factors for many common diseases, the precision medicine approach guides clinicians in therapeutic care and promises to be significantly better with fewer side effects. For example. Scripps Research Institute, in CA, in collaboration with Intel developed an algorithm that identified 23 patients with elevated cardiovascular disease risk that was undetected by traditional statistic methods. This algorithm has 85% accuracy. (4)

Robotics

Robots with built-in microprocessors and sensors are capable of performing a variety of complex human tasks on command by remote control. Artificially intelligent healthcare robots have been designed to assist with patient care such as. measuring blood pressure, heart rate, heartbeat irregularities, and body temperature. They are also used to help in the care of the elderly or chronically ill patients, and those with cognitive decline by reminding patients to drink fluid, to take medicine, or to remember an appointment. They are also used for collecting data, monitoring for emergencies and assisting people with domestic tasks.

By the beginning of this century, robots capable of performing complex assignments could assist human surgeons in performing minimally invasive surgery. Since their precision is extremely accurate, they reduce the number of people needed in the operating room and enhance surgical performance with their superior dexterity.

In VA hospitals today, six-foot tall robots equipped with a fifteen-inch flat screen, two high-resolution cameras, and a microphone engage in two-way video conferencing as they make rounds and see patients. The physician appears on the screen and carries on a conversation with the patient as easily as if he or she were physically present. The doctor remotely drives the robot through the hospital to each patient's room to check vital signs, inspect incisions, and discuss treatment options in real time. In a future scenario of robotic telepresence, it will be commonplace to enable physicians to conduct remote patient rounds, consult with staff, and access patient data via two-way real-time streaming video, wireless protocols and use of the internet. Specialists who may be hundreds of miles away will also use robots to virtually consult with patients. There will be robots that help in the emergency room and other robots that will monitor patients in the intensive care unit.

The Practice of Medicine in 2050 and Beyond

The cornerstone of healthcare practice in 2050 will be that the patient occupies center stage. Interactions between patients and provider will take place whenever and wherever feasible, via face-to-face visits in the doctor's office, at a house call, on a web portal, on a smartphone, in a quick response clinic, or, as a last resort, in the hospital emergency department. All fundamental healthcare services will be provided to individuals by a healthcare gatekeeper, who will be responsible for coordinating care, controlling costs, and communicating results. Advances in telecommunications, medical imaging, massive intelligent databases, robotics, satellite technology, and other information systems will allow physicians to communicate far more easily and quickly, and will enable patients to educate and empower themselves to take responsibility for their health.

Patients who live in urban and suburban centers and who are too elderly or infirm to leave their place of residence will be visited at home by physicians, nurses, nurse practitioners, or physician assistants. These same patients will be monitored daily at telemedicine centers. Portable computing technology, tablets, smartphones, access to patient portals, and the assistance of robots will help to prevent illness from escalating out of control. The task of managing the resources of healing is one of the most complex tasks that society must face. By the second half of the century, a realignment of provider networks into patient-centered care management groups will offer continuous care whenever and wherever you need it. Wireless networks will enable patients to access healthcare assistance in the car, at the office, at the beach, or in the gym. Routine tests and examinations that used to require an in-office visit will be done remotely via connected EKGs, EEGs, and portable telehealth units that include tactile devices in the form of hats, gloves, floor mats, bracelets, and toothbrushes that communicate via your smartphone or tablet.

The overall cost of healthcare will continue to rise due to an aging world population. During the first half of the century, the global population of individuals age sixty or over is projected to expand by more than a factor of three. The Census Bureau forecasts that by 2050, 88.5 million people will be aged sixty-five and older; 19 million people will be aged eighty-five and older; and 601,000 people will be aged one hundred and older (under this scenario, one in every three persons living in the more developed regions of the world are likely to be sixty or older, and about one in every four is projected to be sixty-five or older. (5)

During the last sixty to ninety days of life, there is a disproportionate amount of money spent on an individual's healthcare, which impacts resources available throughout the system. By 2050, a coalition of government, private organizations, employers, hospitals, physicians, payers, and foundations will oversee how we pay for care, who pays what, and how we overcome the inequities in the system. Extraordinary health expenses will come out of an individual's health account set up either with matching grants from employers or with personal funds. Individuals who cannot afford health accounts, who are unemployed, disabled or otherwise unable to work, will be supported by public funds.

In 2050, patients will be offered incentives and will be provided with the tools, such as critical markers for diet, blood pressure, blood sugar, which is their responsibility to keep in check. Adherence to medication, a weight plan and an exercise program; control of tobacco and alcohol intake; and other lifestyle changes will be rewarded with lower rates for healthcare coverage, similar to the penalty/reward system used today by auto insurers. This will become part of the high value that is placed on healthy living. Additionally, a system of online social health networks will support patients by educating them through extraordinary health events.

Privacy issues continue to loom large in healthcare 2050. The more digital data we have, the greater the threat that it can and will be penetrated. Although the points of entry will be tightly controlled with advanced encryption technology and scanning capabilities (e.g., retinal scanning of individuals before accessing a record), we know that data is always subject to invasion and tampering by savvy individuals, companies, or countries controlled by demagogues, who desire to access those records for profit or illicit purposes.

By 2050 medical students, during their training, will learn how to deploy and use sophisticated digital information systems so that graduating physicians will begin their careers prepared for the digital world they will live in with their patients.

This is a snapshot of what health 2050 will look like, and how it will benefit e-Patients as they learn to navigate their way through the digital world to stay healthy and well.

Key Points

1. The availability of information technology that fosters the potential for twenty-four-seven communication between e-health professionals and e-Patients forms the essence of their relationship in Healthcare

2050. Patients are the cornerstone and are served in many ways, from home monitoring to retail clinics to web visits to traditional care in the doctor's office or at the hospital. With their mobile devices, they are an integral part of the healthcare team.

2. A microchip embedded in every individual in 2050, which points to a personal health record residing on a secure site on the internet, resolves the issue of who owns your digital health record and how it is accessed by your providers.

3. Patients in 2050 are educated, motivated, and incentivized to pay attention to their health and work toward a healthy lifestyle, thus curbing some of the demand for healthcare services.

4. Many devices and enablers, including smartphones, tablet computers, RFID, robots, extensive telemedicine networks, and decision support systems, all based on sophisticated wireless technology, foster better, safer medicine for everyone.

5. Basic health coverage is the norm in 2050. Those who are employed receive their care in their benefit package from their employers. They have to save and pay for their extraordinary health expenses, however, which are not as readily available through their health coverage as they had been. Those who are not employed, such as seniors, children, and disabled individuals, have access to healthcare coverage through a collaborative of private and public funds.

6. Physicians are taught in medical school about the logistics and value of information technology, and it becomes part of their everyday professional life.

7. Precision medicine, based on our understanding of the human genome, has changed medicine by 2050, More effective ways to predict individual susceptibility to disease, more useful and person-specific tools for preventing disease, greater ability to detect the onset of disease based on markers, and preempting the progression and severity of disease will make the e-Patients health world a new and better place.

Table 4

Major Trends in Healthcare

Trend	Impact	Challenges
Electronic Health Records	Instant access to information; decision support tools	Privacy of health information
Personal Health Records	Empowered patient	Keeping the record up to date
Smartphones	Instant access to health information	Filtering good apps from bad
Email	24/7 asynchronous communication	Focusing on the issue and keeping the email concise
e-visit	Saves money and time by eliminating unnecessary office visit	Ensuring the interaction is meaningful; e-visiting only with physician you know
e-prescribing	Elimination of medication error, speed in obtaining medication	Medication adherence if reasons for drug are not properly explained
Retail Clinics	Fast and economical way to address minor ailments	Integration with primary care physician
Nurse Practitioners and Physician assistants providing primary care	Addresses shortage of PCPs and enables more people to access basic medical care	Ensuring best practices; determining when to refer to MDs
Telemonitoring	Keeps people in their home	Reimbursement, dependence on patient understanding and cooperation
Payers influence in care decisions	Healthcare driven by finances and not specific medical issues	Maintaining integrity of care
Internet	Information access, collaboration, research, communities	Information overload and quality issues

Notes

1. Use of AI in Healthcare and Medicine is Booming," https://www.insiderintelligence.com/insights/artificial-intelligence-healthcare/

2. https://www.ibm.com/cloud/learn/what-is-artificial-intelligence#:~:text=Artificial%20intelligence%20leverages%20computers%20and,capabilities%20of%20the%20human%20mind.

3. Journal of Medical Research (i-JMR) article, "Information Needs in the Precision Medicine Era: How Genetics Home Reference Can Help." (Available through PubMed Central) NIH⟩, https://medlineplus.gov/genetics/understanding/precisionmedicine/definition/

4. https://www.kolabtree.com/blog/5-real-world-examples-of-ai-in-healthcare/#4_Precision_medicine

5. https://www.ncbi.nlm.nih.gov/pmc/articles/PMC2888016/

GLOSSARY OF TERMS

Adherence: Patients following a prescribed treatment ordered by a physician, including the taking of medications at the times and dosages prescribed.

Agency for Healthcare Research and Quality (AHRQ): Part of the US Department of Health and Human Services. AHRQ supports research to improve the outcomes and quality of healthcare, reduce costs, address patient safety and medical errors, and broaden access to effective services.

The American Recovery and Reinvestment Act of 2009 (ARRA): Includes the Stimulus Act with a provision that provides financial incentives for physicians to implement electronic health records.

Centers for Disease Control and Prevention (CDC): The CDC is a Federal Agency within the Department of Health and Human Services, whose mission is to protect America from health, safety, and security threats, both foreign and in the U.S. CDC enhances and promotes the health of U.S. citizens by monitoring and controlling disease, working with states and other partners to maintain a system of health surveillance and prevention of disease outbreaks, including bioterrorism. CDC implements disease prevention strategies, maintains national health statistics, and provides for immunization services. Among its goals is to oversee workplace safety, and environmental disease prevention, as well as ensuring that the United States has strong, well-resourced public health leaders and capabilities at the national, state, and local levels to protect Americans from health threats.

Centers for Medicare and Medicaid Services (CMS): CMS is the Federal agency within the Department of Health and Human Services that administers Medicare, Medicaid, and the State Children's Health Insurance Program, health benefit plans that engage covered individuals in choosing their own healthcare providers, managing their own health expenses, and improving their own health with respect to factors that they can control.

Community health centers: These are Federally funded medical facilities designed to provide ambulatory health services to underserved populations in both rural and urban areas.

Digital dashboard: A control panel used in the emergency department of a hospital to track and monitor patients.

Digital signature: A mathematical scheme for demonstrating the authenticity of a digital message or document. A valid digital signature gives a recipient reason to believe that the message was created by a known sender and that it was not altered in transit. Digital signatures are commonly used for software distribution, financial transactions, and in situations such as healthcare, where it is important to detect forgery or tampering.

e-health: The delivery of healthcare services that incorporates the use of information technology and fosters a collaborative environment between a patient and a provider, which includes sharing of information and interactive communication.

Electronic Health Record (EHR): An aggregate electronic record of health-related information on an individual that conforms to nationally recognized interoperability standards and that can be created, managed, and consulted by authorized clinicians and staff across more than one healthcare organization. The EHR includes patient demographics, progress notes, problems, medications, medical history, immunizations, laboratory data, and radiology reports and images.

eICU: A location that is removed physically from the intensive care unit in a hospital where physicians can monitor several ICUs at the same time. The eICU can be located within a hospital complex or off-site miles away, and is manned around the clock by intensive care doctors and nurses.

e-Patients: Empowered healthcare consumers who take an active role in their healthcare, collaborate with their providers, and use all available digital communication technology including the internet, email, smartphones, personal health records, and other applications and tools to manage and monitor their health and the health of family members.

Electronic prescribing (e-prescribing): Entering a prescription in a data entry system and generating that prescription electronically.

e-Visit: An asynchronous online discussion between a clinician and a patient via the internet in a secure environment for the purpose of resolving issues and answering questions between office visits.

Electronic mail (email): The exchange of digital messages across the internet.

ED/ER: Emergency department/emergency room in a hospital.

Flexible savings accounts (FSA): An FSA allows an employee to set aside a portion of his or her earnings to pay for qualified expenses most commonly for medical expenses. It is similar to a health savings account (HSA) or a health reimbursement account (HRA). However, while HSAs and HRAs are almost exclusively used as components of a consumer-driven healthcare plan, medical FSAs are commonly offered with more traditional health plans. Paper forms or an FSA debit card, also known as a flex card, may be used to access the account funds.

Food and Drug Administration (FDA): The US Federal Agency responsible for protecting the public health by assuring the safety, efficacy, and security of human and veterinary drugs, biological products, medical devices, food supply, and cosmetics.

Gatekeeper: A professional healthcare provider who oversees and coordinates care for patients. That person could be a primary care physician, nurse practitioner, or physician assistant.

HapMap: Describes the common patterns of human DNA sequence variation. The HapMap database is a key resource for researchers to use to find genes affecting health, disease, and responses to drugs and environmental factors.

Health Information Exchange (HIE): The electronic movement of health-related data and information, including email, the internet, digital databases, audio and video, among organizations according to specific standards, protocols, and other agreed criteria. (Source: NAHIT)

Health Information Technology (HIT): The deployment of digital communication tools, including email, the internet, digital databases, audio, and video, to facilitate information exchange among healthcare providers and patients.

Health Insurance Portability Accountability Act (HIPAA): Legislation enacted in 1996 by the US Congress to protect health insurance coverage for workers and their families when they change or lose their job and to define national standards for electronic healthcare transactions and address the privacy and security of health data.

Health Maintenance Organization (HMO): An organization that provides healthcare coverage in the United States through hospitals, doctors, and other providers with which the HMO has a contract. The HMO covers only care rendered by those doctors and other professionals who have agreed to treat patients in accordance with the HMO's guidelines and restrictions in exchange for a steady stream of customers.

Health practice shortage area (HPSA): A region in the United States where there is a shortage of primary care physicians, dentists, and mental health practitioners as defined by CMS.

Healthcare payers: Insurers, including health plans, self-insured employer plans, and third-party administrators, providing healthcare benefits to enrolled members and reimbursing provider organizations.

Health Reimbursement Accounts (HRA): Funds that a set aside by an employer on behalf of an employee for healthcare needs.

Health Savings Accounts (HSA): Funds set up to assist individuals with payment of out-of-pocket health expenses. Both employers and employees contribute to the HSA up to an annual amount limit set by statutory cap.

Hill-Burton Act of 1946: Provided funds to states for the construction of hospitals.

ICE (in case of emergency): An acronym that denotes an emergency contact who can be reached and who has access to an individual's electronic health record when that individual arrives in the emergency department unconscious and unable to communicate with the doctors.

Institute of Medicine (IOM): was established under an 1863 Congressional Charter as a non-profit institution that would provide objective advice on matters of science, technology and health. Renamed the National Academy of Medicine in July 2015, the mandate of the institution is to improve health for all by advancing science, accelerating health equity, and providing independent, authoritative, and trusted advice nationally and globally.

Medicare: A social insurance program administered by the United States government, providing health insurance coverage to people who are aged sixty-five and over.

Medigap: Refers to private insurance plans sold to Medicare beneficiaries that provide coverage for medical expenses not or only partially covered by Medicare.

National Health Information Network (NHIN): An initiative for the exchange of healthcare information being developed under the auspices of the US Office of the National Coordinator for Health Information Technology (ONC).

Nurse Practitioners: Individuals who are licensed to practice medicine, to see patients, give exams, make diagnoses, make referrals, and treat acute illness, injuries, and infections, as well as prescribe medicine.

ONC: This is an acronym for the Office of the National Coordinator for Health Information Technology under the auspices of the Secretary of the Department of Health and Human Services. The national coordinator serves as the secretary's principal advisor on the development, application, and use of health information technology in an effort to improve the quality, safety, and efficiency of the nation's health through the development of an interoperable harmonized health information infrastructure.

Patient-centered medical home (PCMH): An approach to providing comprehensive primary care that encourages collaboration between individual patients and their personal providers, and when appropriate, the patient's family. The PCMH becomes the main source of care for the patient.

Patient portal: A secure website set up by a medical institution or a clinical practice where patients and doctors can maintain two-way password protected communication and where access to an electronic health record enables the patient/provider team to review medications, medical history, lab results, treatment programs, and other aspects of a patient's care.

Patient Protection and Affordable Care Act (PPACA): The Patient Protection and Affordable Care Act is a federal law that was signed by President Barack Obama on March 23, 2010. This act, along with the Healthcare and Education Reconciliation Act of 2010 (signed into law on March 30, 2010) focuses on reform of the private health insurance market and includes provisions that provide improved coverage for those with preexisting conditions. It also addresses prescription drug coverage in Medicare and extends the life of the Medicare trust fund.

Pay for Performance (P4P): A payment approach used in healthcare that correlates payment to a physician with how well the physician adheres to practice standards and achieves certain outcomes based on a set of performance measures.

Personal Health Record: An electronic cumulative record of health-related information on an individual drawn from multiple sources that is created and managed by the individual. The data in the PHR and control of access to it are the responsibility of the individual.

Pew Internet and American Life Project: An initiative of the Pew Research Center, a nonprofit "fact tank" funded by the Pew charitable trusts founded by Joseph Pew. The Pew Internet and American Life Project provides information on the issues, attitudes, and trends with a particular focus on the impact of the internet on children, families, communities, the workplace, schools, healthcare, and civic/political life.

Physician Assistant (PA): A healthcare professional licensed to practice medicine, with the supervision of a licensed physician, who works with patients in prevention, maintenance, and treatment, particularly individuals with chronic conditions. PAs conduct exams, order and interpret tests, counsel on preventive care, prescribe medicine, and sometimes assist in surgery.

Point of Service plan (POS): A managed-care health insurance system where patients are required to choose a primary care physician from an approved network to monitor their healthcare in exchange for a lower cost for their insurance.

Preferred Provider Organization (PPO): A managed care organization of medical doctors, hospitals, and other healthcare providers who have entered into an agreement with an insurer or a third-party administrator to provide healthcare at reduced rates to the insurers or administrator's clients or patients.

Primary care physician (PCP): A clinician who provides integrated healthcare services and who addresses a large majority of personal healthcare needs of patients in a sustained long-term relationship.

Privacy Rule: Legislation enacted by the US Congress in 2003 to establish the first federal privacy standards to protect patient health information, protected health information: (PHI): Any information that concerns the health status, provision of healthcare, or payment for healthcare that can be linked to an individual.

Provider: Any registered, licensed health professional who administers care to a patient. This can also refer to healthcare delivery organizations.

Registries: Organized systems for the collection, storage, retrieval, analysis, and dissemination of information about individuals to support health needs. This also includes government agencies and professional associations that define, develop, and support registries.

Regional health information organization (RHIO): A health information organization that brings together healthcare stakeholders within a defined geographic area and governs health information exchange among them for the purpose of improving health and care in that community.

Retail clinic: A medical facility, generally located at a retail pharmacy or other walk-in center (no appointment required), where patients can receive basic healthcare services during regular hours as well as evenings and weekends.

Fees for services are generally paid by the patient and are comparable to a standard co-payment.

Radio frequency identification (RFID): Automatic identification based on radio waves, using devices called RFID tags or transponders that include an integrated circuit for storing and processing information and an antenna for receiving and transmitting the signal. RFID is used in healthcare to match patients with medication and to track materials and people within the healthcare setting.

Robots: Mechanical devices that include the software to make them capable of performing a variety of complex tasks such as measuring blood pressure, heart rate, and body temperature or inspecting incisions and performing surgical tasks. These robots are monitored and assisted by a physician.

Specialist: A healthcare provider who is trained in a specific area of medicine and offers care in that particular branch of medicine.

Smart card: Similar to an ATM or credit card, it contains a patient's complete health history, including medications, insurance data, providers' names, and family history.

Smartphone: A mobile phone that includes advanced connectivity and communication functionality, which is based on an operating software system that enables a great variety of applications. Among these applications are full internet access, email connectivity, cameras and video cameras, GPS systems, and audio features that enable the user to dictate notes into the device that can be transmitted to an electronic health record or stored and saved or other storage medium. Other applications include use as an MP3 player, an e-book reader, movie viewer, and more.

Telemedicine: The use of communication equipment to link healthcare practitioners and patients who are in different locations, allowing patients to receive care where and when it is needed. Telemedicine incorporates videoconferencing, the transmission of images, the use of patient portals, remote monitoring of vital signs, continuing medical education, and nursing call centers.

Telerehabilitation/teletherapy: The clinical application of consultative preventative and therapeutics services via interactive telecommunication systems.

Telemonitoring: Home healthcare using various digital devices and noninvasive wearable monitors that measure blood pressure, blood glucose, weight, and heart rate and send these measurements through a telephone system or through a computer to a remote site where medical providers can check the readings and take appropriate action.

ACKNOWLEDGMENTS

Many individuals including physicians, nurses, nurse practitioners, hospital administrators, academics, technologists, commentators, and especially patients contributed thoughts, ideas and input for this book. These individuals helped because they believe in participatory medicine, patient empowerment, and healthcare delivery supported by digital communication technology that enables us to experience healthcare systems that are more efficient, effective, and economical than the medical practices of the paper-based twentieth century.

I owe a special thanks for their personal interest and mentoring of this project to: Dr. Joseph Kvedor Founder and Director of the Center for Connected Health at Partners Healthcare, Dr. Dan Teres, Medical Director at AstraZeneca, and Professor of Medicine at Tufts University Medical School, Dr. Danny Sands, assistant clinical professor of medicine at Harvard Medical School and senior medical informatics director for Cisco Systems Inc., Susannah Fox, Associate Director Digital Strategy Pew Internet and American Life Project, John Glaser, Chief Executive Officer of Siemens Health Services Business Unit. and former CIO of Partners Healthcare, Dr. Dena Pushkin, Director, Federal Office for the Advancement of Telehealth, US Department of Health and Human Services, Michele Garvin, Esq. Ropes and Gray LLP, Dr. William F. Bria, Chief Medical Information Officer at Shiners Hospitals for Children.

Other individuals who contributed invaluable information include:

Susan A. Abookire, MD, MPH, Chair of Quality and Safety Medicine, Mount Auburn Hospital, Cambridge, MA.

Patricia Abbott, PhD, RN, Associate Professor and Co—Director WHO PAHO Collaborating Center for Nursing Knowledge Information Management and Sharing, Johns Hopkins University School of Medicine.

Thomas Abrams, Director, Division of Drug Marketing, Advertising, and Communications (DDMAC). Food and Drug Administration.

Holt Anderson, Executive Director, North Carolina Health Information and Communications Alliance.

Nina Antonetti, Telemedicine Manager, Marshfield Clinic Telehealth.

Dr. Peter Basch, MD, Senior Fellow. Center for American Progress, Director, Med Star Health, General Internist.

John Blair, MD, President & CEO, Taconic IPA.

William F. Bria, CMIO Shriners Hospitals for Children.

William Braithwaite, MD, PhD, FACMI, HISPI Vice Chairman, Health Information Policy Consultant.

Claire Broome MD, Former Director Integrated Health Information Systems, Centers for Disease Control and Prevention, Adjunct Professor Department of Global Health, Emery University School of Public Health.

Todd Brown, Associate Clinical Specialist and Vice Chair, Department of Pharmacy Practice, School of Pharmacy, Northeastern University.

Gary Christopherson, MD, Senior Advisor; Undersecretary for Health, Veterans Administration Senior Fellow, Institute of Medicine.

Homer L. Chin, MD, Medical Director, Clinical Information Systems, Kaiser Permanente.

David C. Classen, MD Vice President, First Consulting Group.

Jeffrey Cooper PhD, Director of biomedical engineering, Partners Healthcare, Boston MA.

Ted Cooper, MD. Associate Clinical Professor of Ophthalmology, Stanford University Medical School

Robert Cox, MD, Director Hays Medical Center, Hays Kansas.

Eliot Cutler, Director, Blood Lab at New England Medical Center, Boston.

Thomas Delbanco, MD Richard Florence Koplow-James Tullis Professor of General Medicine and Primary Care, Harvard Medical School, Founder, Division of General Medicine and, Primary Care Medicine Beth Israel Deaconess Medical Center, Boston, MA.

Suzanne Delbanco, Ph.D. founder and Executive Director of Catalyst for Payment Reform, formerly of The Leapfrog Group.

Henry DePhillips, MD Chief Medical Officer, Medem.

George Demetri, MD, Director of Center for Sarcoma and Bone Oncology Dana Farber Cancer Institute, Associate Professor of Medicine Harvard Medical School.

Don E. Detmer, MD, President and CEO American Medical Informatics Association.

Joyce DuBow, Associate Director, AARP Public Policy.

David Ellis, MD, Associate Clinical Professor of Emergency Medicine, University of Buffalo

Ross D. Fletcher, MD, Chief of Staff VA Medical Center, Washington DC.

Mark Foster MD, Vice Chairman Taconic IPA Inc., Chairman THINC RHIO.

Charles Ganley, M.D., Director. Division of Over-the-Counter Drug Products. FDA.

Janlori Goldman Columbia Health Privacy Project.

John Halamka, MD, MS, Chief Information Officer Caregroup Health System, Chief Information Officer, Harvard Medical School.

Claus Hamann Md. MS FRCP(C) Geriatric Primary Care Massachusetts General Hospital.

Matthew R. Handley, MD, Primary Care, Family Practice, Group Health Cooperative Seattle, WA.

Dan Hoch, MD, Department of Neurology, Massachusetts General Hospital.

Carol Holquist RPh, Director Division of Medication Errors and Technology Support, FDA, Washington, D.C.

David Jacobson, Professor of Anthropology, Brandeis University.

Joel Kahn MD, President, WorldCare Global Health Plan.

David W. Kaplan, MD, MPH, Chief Medical Information Officer, Children's Hospital, and Professor of Pediatrics, University of Colorado School of Medicine.

Charles M. Kilo, MD, MPH, Greenfield Health, Portland OR.

Bernard Koucher, MD, Founder, Doctors without Borders.

Howard M. Landa, MD, Kaiser Permanente Department of Pediatric Urology.

Thomas F. Landholt, MD, Family Practice, Springfield, MO.

David J. Lansky, PhD, President and CEO Pacific Business Group on Health, Former Senior Director Health Programs at the Markle Foundation.

Mark Leavitt MD, PhD Chairman of the Certification Commission for Healthcare. Information Technology (CCHIT).

Eric M. Liederman, MD MPH, Director of Medical Informatics, Kaiser Permanente UC Davis, Health Systems.

Steven R. Levisohn MD, Internal Medicine, Massachusetts General Hospital, Boston, MA.

Janet Marchibroda, Chief Healthcare Officer Global Healthcare and Life Sciences, IBM, Founder, eHealth Initiative.

Robert J. Mandel MD, MBA eHealth Program, Blue Cross Blue Shield, Massachusetts.

Bradford Middleton, MD, MPH, Chair Clinical Informatics, Research and Development, Clinical Informatics Research Department at Partners Healthcare.

David Nash MD, MBA Professor and Chairman of Health Policy, Thomas Jefferson University Medical College, Philadelphia, PA.

Larry Nathanson MD, Director Emergency Medicine Informatics, Beth Israel Deaconess Medical Center, Boston, MA.

Marc Overhage, MD, President and CEO of the Indiana Health Information Exchange.

David Rattner, MD, Chief of Division of Gastrointestinal and General Surgery, Massachusetts General Hospital.

Brian Rosenfeld, MD, Founder VISICU Inc.

David Rose, Chief Executive at Vitality and Founder of Ambient Devices.

Steve Ross, MD, Associate Professor University of Colorado Health Science Center, General Internal Medicine, University of Colorado Hospital.

Jay H. Sanders, M.D. President and CEO of the Global Telemedicine Group, Professor of Medicine at Johns Hopkins University School of Medicine.

Joseph Scherger, MD, MPH, Professor of Clinical Family and Preventive Medicine at the University of California, San Diego School of Medicine.

Steve L. Schneider MD Associate Director, Healthwise, Co-founder Idaho Wellness Center.

Hasan Sharif MD, Founder, COO, CMO, WorldCare.

Linda E. Silvers, Health Fraud Team, US Food and Drug Administration.

Warner Slack, MD, Department Neurology, Beth Israel Deaconess Medical Center, Boston, MA.

Joseph Siemienszuk, MD, CMO Providence Health System.

Paul Tang, MD, MS, Vice President and Chief Medical Information Officer Palo Alto Medical Foundation (PAMF). Stanford CA.

Lisa Vetter Telemedicine Specialist St. Alexius Medical Center North Dakota.

Jonathan S. Wald, MD, MPH Former Associate Director of the Clinical Informatics Research and Development (CIRD) Group Partners Healthcare, Director PC Technology at RTI International.

Kenneth Weiss, PHD. Psychologist and Clinical Instructor, Harvard Medical School.

Andrew Wiesenthal, MD, Associate Executive Director for Clinical Information Support the Permanente Federation.

David Williams Principal, MedPharma Partners LLC.

Eric Zimmerman, Senior Vice President, Relay Health.

Special thanks to my graphic artist, Tania Helhoski of Bird Design and to Ed Shems, a talented illustrator who did the cartoons that appear in the book.

www.ingramcontent.com/pod-product-compliance
Lightning Source LLC
Chambersburg PA
CBHW060459290526
45791CB00001B/186